WATER

Also by Alice Outwater:

The Cartoon Guide to the Environment (with Larry Gonick)

WATER

A NATURAL HISTORY

ALICE OUTWATER

ILLUSTRATIONS BY BILLY BRAUER

BASIC
BOOKS
A Member of the Perseus Books Group

Designed by Elliott Beard

Library of Congress Cataloging-in-Publication Data

Outwater, Alice B.
  Water : a natural history / Alice Outwater ; illustrations by Billy Brauer. — 1st ed.
      p. cm.
  Includes bibliographical references and index.
  ISBN 0-465-03779-8 (cloth)
  ISBN 0-465-03780-1 (paper)
  1. Hydrology—United States. 2. Hydraulic engineering—United States. I. Title.
  GB701.O88   1996
  551.48—dc20   96-24182
                                                          CIP

01 ❖/RRD 10 9 8

Water is printed on recycled paper.

To Bob, without whose unfailing support and encouragement this book would never have been written, and to Sam, who cannot remember a time when his mom wasn't working on *Water*.

CONTENTS

ACKNOWLEDGMENTS

This book would not exist without the help of Larry Gonick, the nonfiction cartoonist from whom I learned the art of deep research and terse writing during our work on *The Cartoon Guide to the Environment*. Vicky Bijur, my agent, showed me step by step how to make a real book; Susan Rabiner, my editor at Basic Books, helped me see what I had to do and never tormented me because of how long it took. Sara Lippincott, my editor in Pasadena, was like a good fairy: she rewove the book into whole cloth.

Thanks to Dr. Penny Chisholm at MIT, who startled me into being far more rigorous, for her review of the material on beavers; and to Dr. George Pinder at the University of Vermont, whose expertise in hydrology was of great assistance. Betty Howlett, of the Joslin Memorial Library in Waitsfield, Vermont, got some hard-to-find books for me, and the staff of the University of Vermont library was unfailingly helpful. Janet Brown provided daily encouragement, Johnny Summers and

Charles Fallick kept me in good humor, and Lise Wexler checked my French translations. My sister Anne helped clarify for me the vision of the land that was; my sister Catherine reminded me of the importance of detail; my brother John provided engineering input and helped me think out the material.

Thanks to Dr. Richard M. Blaney of Brevard Community College, in Florida, for his information about estrogenic chemicals; to Donald L. Ferry, for his clips from cyberspace; to Lance Olsen, for his thoughts on forests; and to Reginald Rockwell, whose ideas about "new journalism" helped shape the last chapters.

Many thanks to my father, Dr. John O. Outwater, who has been integrating engineering concepts and the natural world for a lifetime, and to my mother, Dr. Alice D. Outwater, for keeping me close to beavers. And many thanks also to Elizabeth Stanton, who first encouraged me in this project six years ago, and to Anne Fried, my high school English teacher. I hope she doesn't cringe at all the dashes.

~

INTRODUCTION

In 1972, the United States passed the Clean Water Act in response to a crisis in national water quality. Its purpose was to restore and maintain the chemical, physical, and biological integrity of the waterways: by 1985, the discharge of pollutants into the waterways was to cease, and all of the nation's streams, rivers, and lakes were to be fishable and swimmable. Every city was required to build a secondary wastewater treatment plant, and every industry had to install the best available technology to reduce the discharge of pollutants into the waterways. In the years that followed, the stranglehold that wastes had on the nation's streams, rivers, and lakes was eased.

However, stringent discharge controls have not been enough to restore the nation's waterways: a generation after the Clean Water Act was passed, about a third of the stream miles and lake acres in the United States are still polluted. Obviously, there is a great deal more that we need to do.

This book was born in the bowels of a Boston wastewater treatment plant. I have a bachelor of science degree in mechanical engineering from the University of Vermont, and a master's degree in Technology and Policy from the Massachusetts Institute of Technology, where I concentrated on the ecological, chemical, political, and economic bases of water pollution. When my education was done, in 1987, I went to work on the $6 billion Boston Harbor Clean-Up, looking for land applications for sewage sludge, and there I learned that industries, nationwide, were no longer discharging very much of their wastes to the country's wastewater systems. Because part of my job was to assess sludge quality, I had to track down where the contaminants in sludge were coming from, and in the course of this work I examined municipal sludge reports from all over the country. What I found was that the nation's industries had cleaned up their act. The chemical profile of most municipal sludge looks a lot like the soil and manure that grew your food, with a bunch of toilet paper thrown in. The sludge is smelly all right—but it has very low levels of chemical contaminants.

I had been trained to think about water pollution in terms of the end-of-the-pipe. But for each day that water spends in pipes, it spends perhaps a decade in the natural world. Water is the blood of land—always in motion, from the rain to the mountaintops, through the forests and plains to the sea, and so to the clouds again. And yet, on the North American continent, the natural water cycle has been changed in a number of ways. As a result, water is no longer able to clean itself naturally, and despite our best legislative efforts our waterways are still impaired.

By dredging, by damming, by channeling, by tampering with (and in some cases eliminating) the ecological niches where water cleans itself, we have simplified the pathways that water takes through the American landscape; and we have ended up with dirty water. What we can hope for is that by changing the way we manage our vast public lands—by restoring those elements of the natural world which made the water pristine in the pre-Columbian waterways—we can have clean water once again.

PART I

~

DISMANTLING THE NATURAL SYSTEM

ONE

~

THE FUR TRADE

This story about American water begins, surprisingly, back in Europe during the Middle Ages, when peasants living in cramped quarters close to their animals were probably warmer than the rich in their drafty great halls. Fireplaces, the medieval equivalent of modern space heating, were so inefficient that to keep the winter's chill at bay the well-to-do slept under fur covers, put on fur-lined robes or tunics upon rising, and donned fur-lined cloaks before venturing outside. More furs were worn in the Middle Ages than at any time before or since.

By the end of the thirteenth century, furs had become so much a part of the wardrobe of the times that rich and poor alike wore them in all seasons. Kings and dukes owned as many as twenty or thirty garments lined in fine furs, and would dress in two and even three layers of fur-lined clothing. Records of royal wardrobe purchases by French and English kings list enormous numbers of skins for royal households; from 1285 to 1288, for example, King Edward I bought a hundred and twenty thousand squirrel pelts a year. But the market would have been quite small if only royalty wore furs. Even relatively humble citizens owned one or two robes lined in lambskin or cat which could be bought for a few shillings, or a few weeks' work. In 1363, an English statute stipulated the furs that might be worn by each social class: hoi polloi were restricted to local skins—lamb, rabbit, cat, or fox—while the nobility, clerics, and richer citizens could wear ermine, lynx, sable, beaver, and fine Baltic squirrel.

Not surprisingly, trade in fine furs began early and became quite robust. Beginning in the ninth century A.D., Viking traders collected beaver, sable, and squirrel as tribute from the Finns and exchanged them in England for wheat, honey, wine, and cloth. Skins collected from southern Europe and North Africa appeared regularly at the Marseilles fur markets; those from Scandinavia and Russia were sold at Bruges; Spanish beaver was sold in London. So great was the demand for luxury furs that furriers—called skinners in medieval England—became politically powerful, socially prestigious merchants: the historian Elspeth Veale reports that their mercantile guild was powerful enough to set the length of time for an apprenticeship at fourteen years compared to the usual seven—an arrangement that made skinners among the most exclusive guilds of the merchant class. By the end of the 1300s, the fur business had become so successful and the market so insatiable that most of the wild European furbearing species were gone.

Initially, trappers simply moved deeper into the Russian interior. Complete records for pelts collected from northern and western Russia and shipped from Novgorod to England survive for two periods in the late 1300s: from July to September 1384, a total of 382,982 skins were imported from the Baltics, accounting for 97 percent of England's total fur imports; from March through November 1390, 323,624 furs were imported, of which 96 percent were Russian. In the next two centuries, millions of Russian squirrel, beaver, marten, ermine, and sable pelts were bought by European skinners.

The beaver was in particularly high demand, for a number of reasons. First, its fur is uniquely suited to felting. The undercoat is much finer and denser than lambswool or rabbit hair, and the shaft of each slim hair is covered with tiny barbs that lock together to make the beaver waterproof. When this undercoat was shaved off the skin, pounded, stiffened, and rolled into felt for a hat, the tiny barbs on each individual hair linked the whole so securely that the hat kept its shape far longer than hats made from other furs. Second, the beaver could be used in its entirety. Castoreum, a glandular secretion that oils the beaver's hair and is used to mark its domain, was highly sought after, both as a medicine and for use as a perfume base. Beaver flesh was preferred to that of game birds, and beaver tail was considered a great delicacy; moreover, because of the scales on its tail, beaver flesh could be eaten in lieu of fish during the Lenten season. And finally, since beavers build dams, stay by their pond, and take time to raise their kits, they are easily caught.

Beavers once lived throughout the European wilderness and the British Isles. As people pushed back the forests, and as furriers paid good money for beaver hides, the beavers began to disappear. Scotland's beaver trade dried up by 1350, by which time a beaver pelt could cost up to a hundred and twenty times as much as a lambskin. By then, the Continental beaver was scarce: of 377,200 furs delivered to England from the Baltics in 1384, only 3,926 were beaver. The wardrobe records of France's kings further reflect the devastation of the European beaver population. In eighteen months of 1387 and 1388, King Philippe and his brother were able to order 450 beaver hats for their personal use; the next year the King bought only 144 beaver hats for himself and his son; in 1390, his purchases of beaver hats fell to 62, and he filled out his wardrobe with felt hats made from black lambswool, marten, and rabbit. Year by year, the King of France was able to buy fewer and fewer beaver hats. After 1415, of an annual purchase of hundreds of hats, rarely more than a dozen were of beaver. All the evidence suggests that beaver hats were still prized and that the King's diminishing purchases were not of his choosing. Indeed, beaver hats were so precious that they were often willed to heirs.

Such sixteenth-century innovations as glass windows, better chimneys, and the use of coal as fuel had made it warmer indoors, and fashion changed along with the interior temperature. Sumptuous fabrics and jewels supplanted many of the furs as prized possessions, and the trade in pelts declined. While the reprieve allowed relict European populations of many furbearing species to rebuild, the beaver was simply gone from most of the Continent. By the mid–1500s, only the remote reaches of Siberia and Scandinavia had ponds still abundant in beavers.

But the desire for beaver hats had become an imperative of fashion that scarcity could not destroy. Now, however, there was only one place on earth where millions upon millions of beavers still existed. All that remained was for the last stronghold of the beaver to be discovered.

When the Europeans landed in the New World, the Indians who greeted them were using flint knives, bone awls, and stone or skin kettles. European iron tools were so vastly superior to these Stone Age implements that as soon as the Indians became aware of their existence, they wanted them. Indian women especially welcomed the European technology: iron kettles, knives, awls, and hoes were clearly superior to their own tools, and the woven goods—blankets and coarse cloth—were pleasing alternatives to animal skins. Axes, guns, powder and shot, traps, and access to a market economy would soon change the lives of the men as well. The only product the Indians could exchange for the European goods they coveted were the coats of the animals that crowded the wilderness. And so, in a Faustian bargain of enormous reach—one that would eventually leave them without the means to earn a living and no way to return to the lives they had led before the arrival of the Europeans—the Indians turned wholeheartedly to the task of trapping the North American furbearers.

Beavers were the first to go. By the mid–1600s, beaver hats were once again plentiful in Europe and were being worn by both sexes. A gentleman's attire included a black beaver hat adorned with an ostrich plume; men swept such hats from their heads when they bowed, with an elegant, practiced gesture. In 1638, Charles I of England decreed that "Nothing but beaver stuff or beaver wool shall be used in the making of hats." Samuel Pepys noted in his diary in 1660 that his beaver hat cost £4 5s. (more than a wig and less than a cloak), while a 1719 treatise by the Countess of Wilton describes a fashionable lady's annual wardrobe as including "a beaver and feather for the forest," at a cost of £3—the same price as a set of stays or English shoes. The high-crowned Spanish beaver hat was followed by the conical beaver of the Puritans, which was itself superseded by the broad-brimmed slouch hat of the Restoration, the plumed "shovel" hat of the French Revolution, and the flat-crowned clerical and three-cornered cocked hat that became standard headgear for men throughout the eighteenth century. Everyone wore hats, and the best of these were made from New World beaver.

The effects in Europe of the beaver trade were apparent, but the effects in the New World were far less obvious. In one sense, Europe's insatiable demand for beaver hats did make a noticeable change, allowing the Native American to pass from the Stone Age to the Iron Age in a single bound. The Indians embraced the new tools and goods so quickly that only the earliest white visitors saw Indians in their original state. In 1620, the Pilgrims noted that the Indian villages of Cape Cod were stocked with European kettles and hatchets—presumably bought with beaver skins. "Nor could it be imagined," wrote the colonial historian Edward Johnson in 1653, "that this Wilderness should turn a mart for Merchants in so short a space, Holland, France, Spain and Portugal coming hither for trade." But in fact this is precisely what happened. Stretching over three hundred years, the North American fur trade would alter the physical landscape of the New World as no other trade has done before or since.

At the onset of the fur trade, ten good beavers—adult, winter-prime northern hides that were stretched and cured—bought the Indians one gun. One good beaver bought, variously, half a pound of powder, four pounds of shot, a hatchet, eight jack-knives, half a pound of beads, a good coat, or a pound of tobacco. "The beaver does everything perfectly well," an Indian trapper told a Jesuit priest in 1657. "It makes kettles, hatchets, swords, knives, bread. In short, it makes everything. The English have no sense—they give us two knives for one beaver skin."

Neither side understood that another exchange, far more deadly in nature, was simultaneously occurring. It is thought by many that Columbus brought syphilis—endemic in the Americas—back to Spain. What is known is that within a year of his return the disease had entered France. When the young French King Charles VIII led his army of thirty thousand men against Naples in 1494, the ranks were filled with soldiers hired from across Europe. In 1495, the unsuccessful campaign was over, and Charles's mercenaries returned to their homelands with contagious genital lesions. A victim could be dead in two weeks of a pustular rash that ulcerated down to the bone, or survive for decades while the neurological ravages of the disease progressed from blindness to madness and death. Millions of Europeans died from syphilis in the 1500s, and the pandemic taught people of another time the hard-earned lesson that casual sex carried the threat of deadly disease. Syphilis eventually helped to transform the licentious Old World libertines into Puritans and Pilgrims.

Across the Atlantic, an even more devastating calamity was unfolding, as European diseases emptied out Indian villages. Smallpox, measles, tuberculosis, influenza, and typhus entered the Americas and struck the immunologically unprepared Indians with terrible ferocity in wave upon wave of epidemics. Mortality rates in the initial onslaughts were rarely less than 80 or 90 percent of a village's population. Traditional healing practices were useless against the biological assault, and the Indians quickly learned that the only way to escape the new diseases was to abandon their villages and cast aside family and community ties.

The settled, populous agricultural tribes fared the worst: the people of villages attacked by a new pathogen often missed key stages in their annual subsistence cycle—corn planting, say, or the fall hunt—and were consequently weakened when the next infection arrived. Those who survived, wrote Robert Cushman, a contemporary chronicler, "have their courage much abated, and their countenance dejected, and they seem as a people afrighted." To the Puritans, the epidemics were manifestly a sign of God's providence, a way of making room for them in the new land. More than fifty of the earliest European settlements were sited near abandoned Indian villages, with the settlers appropriating Indian fields. "God hath hereby cleared our title to this place," wrote Governor John Winthrop of the Massachusetts Bay Colony in a letter to his friend, Sir Simond D'Ewes.

Left without any response to these plagues but flight, the Indians concluded that fur trading was their best hope for survival, a decision of great benefit to the Europeans. With the Indians doing the trapping, beaver skins became the coin with which the colonists paid off the debts incurred in establishing a colony, bought additional necessities from Europe, and ultimately acquired modest fortunes. Beaver skins bankrolled most of the early colonists, including the Pilgrims. Captain John Smith had arrived off New England in 1614 "to take Whales and make tryalls of a Myne of Gold and Copper. If those failed, Fish and Furres was then our refuge." He found no gold and caught no whales, so devoted himself to fishing and trading for furs. As his men fished, he ranged up the coast with a small boat and crew and "got for trifles neer 1100 Bever skins, 100 Martins and neer as many Otters; and the most of them within a distance of 20 leagues." It was Smith's assessment of the likelihood of quick returns from the fish and fur trade that convinced a consortium of English merchants to finance a Pilgrim colony in the New World.

When these Pilgrims landed on the tip of Cape Cod in early November of 1620, they spent a month searching for the best spot to build a settlement. Previous voyages had described thriving Indian agricultural villages on Cape Cod and northward, with neat wigwams and well-kept fields of corn, squash, and beans. Instead, the Pilgrims found villages that were deserted but often held evidence of recent human occupation: fields of plant stubble, empty wigwams, caches, burial sites, and European tools. In some villages, skulls and bones were strewn about the ground, suggesting a plague so deadly that no one had survived to bury the corpses. Fields that the Indians no longer had the heart or the numbers to plant were lying fallow, and the Pilgrims happily moved in.

By the time Squanto wandered into their midst the next spring, over half of the Pilgrims themselves had died—not of plague but of malnutrition and too many bad colds. Squanto claimed to be the sole survivor of the Pawtuxet tribe, whose fields the Pilgrims had appropriated. He gave them corn and showed them how to plant it, and when the fields were tilled, the Pilgrims cast about for furs to send back to England to pay off their debt. Volunteering his services, Squanto guided the Pilgrims up to the present site of Boston harbor, on a "voyage to the Massachusetts." They left their boat where Charlestown is now and walked in the direction of Medford, where they met some Indians decked in beaver skins. One member of the expedition reported that Squanto was impatient to "rifle the savage women" of their beaver coats, but the Pilgrims wisely insisted on fair trade. The women "sold their coats from their backs, and tied boughs about them, but with great shamefacedness (for indeed they are more modest than some of our English women are)."

The coats the Pilgrims bought were made of beaver taken in the winter, when the fur was prime. To make such coats, women would scrape the inner side of the pelts and rub them with marrow; each pelt would be trimmed into a rectangle, and from five to eight skins would be sewn together with moose sinews into robes that were worn with the fur side in. After fifteen to eighteen months of wear, the skins became well greased, pliable, and yellow, the fur downy, or *cotonné*, and ready for felting. In the early years of the beaver trade, the courtiers of Europe wore hats that were made from the used coats of North American Indians.

The Pilgrims filled their boat with a "good quantity of beaver" and promised to return later. True to their word, they returned the following March with Captain Miles Standish, and again they had a "good store of trade." Plymouth Colony, with no large navigable rivers nearby, was poorly situated for extensive fur trade into the interior of New England, so the Pilgrims sailed up the coast to trade for furs and extended their fur-trading venture to Maine as early as 1625, bringing home, according to Plymouth Governor William Bradford, "seven hundred pounds of beaver besides some other furs."

The first two shiploads of beaver skins sent to England were captured by pirates, but in 1628 a cargo worth £659 arrived safely, and the Plymouth fur trade grew rapidly. During the period 1631–1636, Governor Bradford estimated that the sales of beaver came to £10,000, a "great sume of money" for such a small colony as Plymouth.

The Plymouth traders were never without competitors in their quest for Maine beaver. In 1620, six or seven ships were trading for Maine furs, and four years later some forty ships from the west of England plied the Maine waters and stopped on shore, where local Indians brought the furs they had gathered during the winter. Many of the towns in Maine, including Augusta, Brunswick, and Portland, were founded as trading posts for beaver skins.

In the Connecticut River Valley, the English entrepreneur William Pynchon was granted a monopoly on the fur trade. By the mid 1670s, nearly a quarter of a million beaver had been shipped to London from the Connecticut River Valley alone, and there were no more beaver to be found in the area.

The Dutch monopolized much of the fur trade south of Cape Cod. In 1624, the Dutch West India Company sent four hundred beaver skins from New Amsterdam. Hudson beaver was plentiful, and tens of thousands of beaver skins were shipped out each year. In 1664 the British colonies captured the city, renamed it New York, and took over the beaver trade. Two decades later, the British complained that "this year . . . the revenue is much diminished, for in other years we used to ship off for England 35 or 40,000 beavers, besides peltry; this year only 9,000." Beaver skins were the principal export of New York until 1700, when the trade ended abruptly. By that time, London had imported nearly two million beaver skins from New York. Since France, Holland, Spain, and Portugal were also trading for American beaver, the total number harvested must have been enormous. Beaver populations that had been stable for thousands of years were effectively exterminated along the American east coast by 1700.

In contrast to the free-for-all in what became the United States, the Canadian fur trade was managed as a sustainable enterprise. Chartered in 1682, the Hudson's Bay Company sought to preserve the furbearing populations. An interval of several years was required between seasonal hunts in a single area, so that populations could reestablish themselves after an area had been trapped out. With exclusive jurisdiction over an immense tract of wilderness, the Hudson's Bay Company enjoyed a monopoly that lasted for nearly two centuries and effectively preserved the Canadian beaver populations. The colonists to the south were not so prudent.

Across the entire sweep of North America, the fur trade was the cutting edge of the frontier, the driving force behind the exploration of the country, and for a time the traders were able to coexist with the Indians. Fur traders generally worked within the existing Indian system, and laid no claim to Indian lands. They took Indian women as wives from the start, and these unions created alliances and cultural bridges between the Indians and the traders. As long as Indians did the trapping, the overhead costs of the fur trade were low, and most tribes welcomed the trade goods as payment. The trader managed the production system, and the Indian provided the labor: white men made fortunes, Indians acquired metal implements, and the fur trade was widely seen as a civilizing agent that introduced the Indians to the concepts of private property and regular work.

The problem with this arrangement was that although the demand for beaver skins was virtually limitless, the supply of beavers was finite. Once tribes had become dependent on trade goods, they were caught in a bind: iron tools break, guns need continuing purchases of powder and shot, cloth and beads would always need replacement. Before the traders came, the Indians had hunted beaver with spears, taking only a few for food and clothing. Now they were killing every beaver they found; whole colonies were destroyed in a season. When the beaver was eradicated from one area, the traders would move on. In 1634, a Jesuit missionary at an outpost on the Great Lakes wrote, "These animals are more prolific than our sheep, the females bearing as many as five or six a year. But when the savages find a lodge of them they will kill all, great and small,

male and female. There is danger that they will finally extermi-
nate the species in this region as has happened among the
Hurons who have not a single beaver, going elsewhere to buy
their skins." His pessimism was prophetic.

In the 1700s, the British colonies occupied a thin slice of
coastland from Maine to Florida, barely 400 miles wide. The
rest of the continent was Indian country, claimed by Spain,
France, and England, but populated with dozens of tribes,
which were no longer so welcoming to the white man. Scalp-
ing and murder were regular occurrences. In some tribes, tor-
ture was a highly refined art and victims would be flayed alive,
painstakingly dismembered, or burned by inches over a period
of days. Treaty by treaty, tribe by tribe, the French, Spanish, and
English created alliances that allowed them to establish posts
farther and farther inland to trade for beaver. The Western fur
grab began in earnest.

Beaver from the interior was shipped out of Montreal, a collection point for skins from the entire Canadian wilderness and
from the upper Mississippi and the Missouri Rivers as well.
There was also serious penetration of white men into the Plains
country of the Sioux. The sons and grandsons of the original
explorers—traders who spoke Indian languages and understood Indian customs—told the Indians tales of great ware-
houses stuffed full of goods. Guided by these white traders,
Indians paddled loaded canoes from the beaver fields of the
upper Mississippi and the Missouri through the Great Lakes to
Montreal, a round trip of over 4,000 miles. Ever lengthening
strings of canoes made this trip, the paddlers sweating, singing,
cursing against the power of the rapids and the bite of the
portage straps.

After the exploration of the Mississippi and Missouri by La
Salle and other French adventurers, dozens of expeditions
ranged up and down the banks of these rivers, armed to the hilt
and carrying caches of trade goods. Instead of paddling laden
canoes from the midcontinental prairies to Montreal, the Sioux

could now trade closer to home. In early 1748, thirty-three Frenchmen were trading with the Comanches, the Wichitas, and the Assiniboins up the Missouri and into the South Platte; the Spanish were trading with the Osages and the Panis-Noirs, the Niobrara and the Arikara; and French treaties with the Comanches and the Jumanos allowed traders into New Mexico. St. Louis, founded in 1763, quickly became the fur-trading capital of the world. Pirogues laden with trade goods would work their way up from New Orleans and return with high piles of furs and robes, often towing great rafts, like high floating islands, carrying still more furs. A single expedition could bring as many as thirty thousand beavers to market.

By the close of the eighteenth century, the beaver was nearly gone, with the last beaver colonies keeping to rough, less accessible country. Moreover, the Indians, pauperized and debilitated by disease, had turned to onshore pirating, and traders traveling through the now fur-impoverished Great Lakes region would be set upon for whatever goods they were carrying. Trappers spent months struggling up the sullen Missouri, then on over buffalo country, where streams were few and far between, through badlands, and into the foothills of the Rockies, where on many of the rocky highland streams they found no scrap of fur in sight. The beaver had been all but eliminated by the colonists in the East, the Hudson's Bay men in the North, and the French and the Spanish in the South and West.

When Meriwether Lewis and William Clark set off to explore the country in 1804, a great part of their mission was to find an artery for the fur trade of the trans-Missouri West. President Jefferson had instructed Lewis to make note of the numbers and species of furbearing animals, to ascertain the attitudes of the Indians to the fur trade, and to establish "the most direct and practicable water communication across the continent for the purposes of commerce." The explorers were heroes, and their travel reports of the northern Great Plains and the Rocky Mountains were widely published in newspapers.

Lewis and Clark wrote of Blackfoot country that was "richer in beaver and otter than any on earth," and of the land of the Crow, where beaver were so plentiful that they could be taken from streams with a club. The river systems that flowed where the Great Plains butt up against the Rocky Mountains were the last great beaver fields found in the United States. When Lewis and Clark returned in 1806, fur-trading expeditions were already saddling up.

After five hundred years of development, the fur trade was poised to take maximum advantage of the last of the beaver stocks. Pack trains of eighty horses would scale the mountains carrying goods, and each company set up its own rendezvous points, where as many as two thousand Indian trappers spent a fortnight camping, drinking, gaming, and bartering their pelts to exchange for tools, cloth, and tobacco.

By the 1820s, the Indians of the Rockies had seen that trapping for trade goods led to cultural disintegration, and they were no longer willing to gather furs for the traders. Without native labor to trap the beaver and transport the skins to posts, the fur companies sponsored white trappers to harvest the furs. For almost two decades, no fewer than a thousand mountain men were working the waterways of the Rockies at any given time. About a dozen large fur companies traded for beaver, and hundreds of smaller, independent enterprises were established, flourished, and faded. (One notable fortune made at that time was John Jacob Astor's, whose American Fur Company shipped beaver skins from the Columbia River Basin . . . so you might say that the beaver built the New York Public Library.) Altogether, the mountain men did their level best, in the face of hostile conditions, to clear out the country's remaining beavers.

By the time the first wagon train made its way West on the Oregon Trail in 1843, the beaver was gone. The mountain men had retired from trapping and were hiring themselves out as guides, and the beaver felt used for the tall top hats had been replaced by silk brought from China. The change in fashion from the dull black beaver hat to the shiny silk topper signaled the end of an era.

With the incentive of European trade and the aid of European tools, the Native Americans had done a thorough job of exterminating the beaver. The tribes got their trade goods, the traders got rich, and the Europeans got their hats. Society has always followed fashion, and to this whim much of the New World lost an animal that shapes the waterways. The beavers had created and maintained an ecological system that enriched the land, but by the time the hat of beaver felt gave way to the silk top hat of the 1840s, the beaver was nearly extinct—so thoroughly trapped in the continental United States that its numbers are still sparse in much of the arid West.

Today, there are between 7 million and 12 million beavers in the United States, most of them around the Great Lakes and along what remains of the Mississippi floodplains. Entire ecosystems based on the beaver's compulsion to spend its nights building dams were slowly but steadily undermined. It would be centuries before the nation figured out exactly what it had lost.

TWO

~

NATURE'S HYDROLOGISTS

The beaver is utterly familiar. Forty inches long and over a foot upright, a beaver seems like a little person with a fondness for engineering. Good-natured, gentle, and clean, it makes a friendly pet that follows its owner around much like a dog, scrambling up onto a lap to be rubbed on the belly whenever it's invited. Beavers were commonly kept as pets around Indian encampments, but they do have a fatal flaw in a modern household: they never stop building. When kept indoors, they will cut down the legs of tables and chairs and build little dams between pieces of furniture. Left on their own, they will rearrange waterways.

Beavers do more to shape their landscape than any other mammal except for human beings, and their ancestors were building dams ten million years ago. These Miocene beavers were 7 feet long, felling trees ages before the mammoths roamed. Their underground spiral burrows can be found from western Europe to central Asia and North America; after their extinction, some of these burrows filled with debris that fossilized, creating twisted masses of stone that geologists call devil's corkscrews. Legends of these prehistoric giants were once widespread. The Indians of Nova Scotia claimed to know of an ancient beaver dam so vast that it flooded the Annapolis Valley; farther west, tales circulated of tribal ancestors using immense beaver teeth to hollow out their canoes.

In tribes across North America, legend had it that the beaver helped the Great Spirit build the land, make the seas, and fill both well with animals and people: Long, long ago when the Great Waters surged in a blind and shoreless world, the gigantic beaver swam and dove and spoke with the Great Spirit. The two of them brought up all the mud they could carry, digging out the caves and canyons and shaping the mud into hills and dales, making mountains where cataracts plunged and sang. Some tribes believed that thunder was caused by the great beaver slapping his tail.

Until European colonization of the New World, *Castor canadensis*, the North American beaver, was one of the most successful mammals on the continent, living almost everywhere there was water, from the Arctic tundra to the deserts of northern Mexico. It was scarce only in the swamps of Florida and Louisiana, where its dams and lodges were no match for voracious alligators. Everywhere else, along thousands of streams, lived colony after colony, dam after dam of beavers in close succession, as many as three hundred dams per square mile, each with its own ring of wetlands.

It is estimated that as many as two hundred million beavers once lived in the continental United States, their dams making meadows out of forests, their wetlands slowly capturing silt. The result of the beaver's engineering was a remarkably uniform buildup of organic material in the valleys, a checkerboard of meadows through the woodlands, and a great deal of edge, that fruitful zone where natural communities meet. Beavers are a keystone species, for where beavers build dams the wetlands spread out behind them, providing home and food for dozens of species, from migrating ducks to moose, from fish to frogs to great blue herons.

A total vegetarian, the beaver eats roots, tubers, and the inner bark of trees. As a consequence, its own meat is sweet and tender, and to avoid becoming everyone's favorite prey it took to the water long ago and is well adapted to aquatic life. Its dense coat conserves heat, and its multipurpose scaly tail functions as a rudder, as a place to store fat for lean times, as an internal temperature regulator, and as an early warning system to other beavers thanks to the noise it makes when slapped against the water.

Served at medieval banquets as "bear's paws," the tail of the beaver covers an even more singular feature. Beavers have neither external testicles nor penis, hence their name—*castor*, from the Latin *castratum*.(Some Victorian references claim that *castor* is derived from *gaster*—belly, in Greek—but that was a shyer era.) The beaver's sexual organs are modestly tucked up inside its body, while a pair of glands in the anal area of both sexes secrete castoreum, the musky oil the beaver uses to grease its coat and mark scent mounds to delineate its territory. Castoreum was a popular medicine in the Middle Ages, said to cure ailments ranging from headaches to impotence; it is high in salicylic acid—the basic ingredient of aspirin—which the beaver ingests by dining on willow bark. Long used as a base for perfume, its scent is described as a pungent, waxy, burnt-orange odor, with smoky notes of Irish peat fires and good pipe tobacco and undertones of cardamom and tea.

Castor's teeth never stop growing; the pair on the upper and lower jaws form curved blades that chisel through wood as hard as rock maple and are perfectly designed for felling trees. Beavers are the largest rodent in North America: at close to sixty pounds, the adult females slightly outweigh the males. The beaver's rotund belly is filled with an enormous gut packed with vegetative matter and the bacteria that convert vegetation into calories. To extract the most calories from its high-fiber diet, the beaver eats everything twice when food supplies are low, a practice called coecotrophy. Ruminants manage this by burping up their ingested food and chewing the cud, but the beaver actually passes food through its entire digestive tract twice, by eating the gelatinous, porridgelike substance that comes out its anus the first time through. Double-digested beaver stool looks almost like pure sawdust.

When they are about three years old, beaver kits leave home to find a companion, with whom they mate for life. During this quest for new territory, they are at their most vulnerable to predators (and today to cars), but they are remarkably safe once they've built their living quarters. When the new couple finds a suitable stream, they mark the area with scent mounds and dig out a den in the stream bank. Beavers are burrowers, and they have powerful, curved claws on all four feet. They're also equipped with a number of features that aid underwater construction: valves close off their nose and ears; thin membranes over their eyes serve as goggles; and skin flaps behind their front teeth allow them to tow tree branches in their teeth without swallowing half the pond. With these adaptations, beavers are able to dig their burrow's entrance well below the surface of the water. Slanting the tunnel upward to above the high-water line, they clear out a room three feet wide and line it well with shredded wood and grass. The underwater entrance keeps them safe from lynx and wolverines, and to ensure that the water stays high enough to hide their burrow's mouth during the low summer flow, they build a dam.

Choosing a dam site where the stream is not too deep and the bottom muck is firm, they fell saplings first and then larger trees. Working by night—sometimes on separate trees, sometimes together on a single trunk—they sit with their paws around the tree, their tails either folded beneath them like a seat or extending behind as a prop. Tilting their heads from side to side, they make deep bites in the tree, driving their long yellow teeth into the wood to wedge, pry, or pull out a chip, chiseling the trunk until the tree topples. After cutting the tree into manageable lengths, they push and pull the logs into position on the dam, pointing the butt ends upstream, and hold them fast with piled mud and stones. As the dam grows higher, the water slows, and the beavers weave in more branches and pat on more mortar until a substantial barrier is completed.

Dams must be continuously maintained, and beavers do so every night, replacing shifted sticks and poles and patting on more mud. They build dams throughout their territory: some for water control; some, it seems, just for fun. A family of beavers can build a 35-foot-long dam in a week.

Where the streams have clearly cut banks and a channel with a uniform current, beavers build a solid bank dam with the poles underneath and earth on top; water discharges through an opening in the dam's crest. If the stream is wide, they bow the dam into the flow of the water, increasing the structure's stability. When the young trees nearby are all consumed and the edge of the forest is too far away for the beavers to drag their branches easily, they dig canals about 2 feet wide and 1 foot deep in which to float the branches back to home base, sometimes extending these canals for hundreds of feet to reach new trees. If a riverbank is steep, they build slides down to the water. Dams more than 4,000 feet long have been found, built by generations of beavers, and nineteenth-century reports describe dams encrusted with lime and half petrified, attesting to hundreds of years of continuous repair.

The beaver is a clever engineer, but its brain is embarrassingly small—smooth and unconvoluted except for the well-developed olfactory lobe. The beaver's ratio of brain size to body weight is the lowest found among mammals: like that of the primitive marsupials, the beaver's brain is about a third the size of the average mammal's; a beaver-size human would have a brain fifteen times as large as a beaver's. Beavers don't have much gray matter, and they don't see well. Nevertheless, there is abundant evidence—noted in Enos Mills's 1913 study *In Beaver World* and confirmed by the contemporary naturalist Hope Ryden, who studied a colony of beavers in New York State—that much of their building technique appears to be learned during their long childhood. Oddly, although the European beaver (*Castor fiber*) is nearly identical in appearance to the American beaver, it has no interest in dam construction; in most regions, European beavers confine their efforts to digging burrows in the stream bank. It seems likely that the fine points of dam construction were lost to *Castor fiber* during the centuries when only a few survived in parks.

Virtually every *Castor canadensis* builds dams, however, and behind each dam the water slowly backs up and covers the land. A rush of insects, animals, and plants transforms that thin sheet of water into a place where every level, every nook and cranny, is teeming with life. Ecologically, wetlands are an example of an ecotone—a transition between two diverse communities. Uniquely, an ecotone contains organisms native to each overlapping community as well as organisms characteristic solely of the ecotone itself. The so-called edge effect—the increased variety and density at community junctions—is what makes wetlands so productive of life, and the beaver's role in this system is to build the dams that make wetlands, increasing the edge between waterways and dry land.

In a wetland, the food web is dense and the niches are varied. Frogs twang in the evening, warning of a raccoon wading out to dig up grubs and insects. Herons stalk the frogs, and migrating ducks settle out of the sky to rest and feast before traveling on. Meadowlarks and magpies alight upon the stumps, and muskrats, voles, and otters make their homes along the shore. Sometimes a moose or a deer wades into the water to eat the greens along the shore, while minnows hide among the stalks.

The crush of insects, animals, and plants in the still water is directly attributable to the lowly algae, the primary producers at the base of the food web. Every tablespoon of wetlands water is crowded with millions of organisms that make up a highly diverse planktonic community, and also with planktonic secretions and excretions, feces, and corpses, along with the debris that washes into suspension from the surrounding land. The plankton include—besides the algae, or phytoplankton—zooplankton and bacteria; these are respectively the plants, animals, and scavengers of this small kingdom.

The tiniest plankton are the scavenger planktonic bacteria, which live freely suspended as single cells or form colonies around a nucleus of dead organic material. They, along with aquatic fungi (the hyphomycetes), clean up the corpses and wastes and other organic debris in the water. About three hundred species of aquatic fungi have been identified: most are tiny structures with four arms; some are curving threads. The fungi and the bacteria are saprophytes—eaters of the dead. The algae, by contrast, transform sunshine and inorganic nutrients in the water into food in order to reproduce. If the largest planktonic bacteria were the size of marbles, the phytoplankton would range from marble size to the size of watermelons, and their colonies would be as large as an elephant.

The algae comprise more than a dozen phyla, in a wide range of shapes: diatoms are built like a tiny box, each enclosing a drop of living material; haptophyta take that basic idea and add two flagella; some phytoplankton are shaped like needles with curved ends; others have spines and projections. The bacteria, fungi, and phytoplankton all feed on impurities, cleaning the water that flows through the wetlands. And can they reproduce! Phytoplankton and bacteria simply divide, in generation times of hours to days. When the sun is shining and the water is warm, phytoplankton divide many times a week, producing trillions of offspring a season.

Phytoplankton can be seen as analogous to the grass that grows in the meadows, and zooplankton—protozoans, rotifers, and tiny crustaceans—as the beasts that graze on them. If phytoplankton were as large as watermelons, with elephantine colonies, the largest zooplankton would be as big as a barn. Still too small to be seen with the naked eye—or, at most, barely visible—these hungry little creatures are made up of little more than an appetite and the means to satisfy it. Pushing their way through the water, some zooplankton wave everything in their reach straight down their gullet. Some spin like tops as they drift along, while others glue themselves to the pond bottom and ingest their prey from above. Protozoans drift like water balloons, undulating near the bottom and engulfing any likely-looking algae they meet, while the Daphnia, a minuscule crustacean, singles out its prey and darts in pursuit of it. Thinking—if they can be said to think at all—only of their next meal, these smallest of animals batten on phytoplankton and on each other, while insects and fish fry eye them hungrily in turn.

The web of life in a wetlands provides something for every appetite. Zooplankton make up the twelve-course meals that the insects eat. The backswimmers, striders, and waterboatmen, the ravenous diving beetles and sly water scorpions, the whirl-a-gigs and coiled mosquito larvae come to eat the grazers, and are themselves food for the fish, amphibians, and birds.

The wetlands' underwater world does a remarkable job of cleaning the water. Not only do planktonic bacteria consume the water's organic contaminants while trillions of tiny phyto-plankton use the inorganics to make food, but the water is cleaned by sedimentation as well. When muddy water from streams and rivers rushes into the stillness of the wetlands, the silt in the water adheres to the stalks of water plants and settles to the bottom. Wetlands clarify water and prevent the soil from washing downstream, and in this fashion fertile meadows are built up. The area where a stream once ran becomes covered with a rich blanket of organic matter.

When beavers make a series of dams and ponds within a drainage basin, the water cycle in the entire watershed is affected. Wetlands act like a sponge, soaking up water during storms and releasing it slowly in drier times. The wetlands that ring a beaver colony's dams and ponds can hold millions of gal-lons of water, and when streams are swollen and muddy with melted spring snow, wetlands reduce flooding and erosion downstream by absorbing much of the excess flow. By provid-ing a flexible reservoir, wetlands reduce flooding from summer storms as well.

Rain that falls on a watershed takes one of four pathways: it evaporates directly back to the air; it flows overland into a stream as runoff; it soaks into the ground, where it's taken up by plants and trees for evapotranspiration; or it seeps down into the groundwater. When water flows as a stream, its motion is turbulent, and soil is carried along with the flow. Swift streams are relatively barren of life, since most organisms are swept away by the force of the current: some flatbodied insects and larvae cling under the rocks, and a few fish eat them. But when a beaver builds a dam and the stream flows into a pond, the water becomes quiescent and warmer, plankton populations increase, and the food base expands. Water detained in the wetlands behind a beaver dam is more likely to percolate down to the groundwater, raising the water table and creating springs and freshets throughout the watershed. A land with hundreds of millions of beavers is a truly rich land, and the wetlands associated with beaver dams made the New World's water plentiful and clear as the dew.

The summer after they build their first dam (having wintered in their streamside burrow while the pond grew), a beaver pair constructs a lodge in the lily pads near the shore. They build a platform of interlaced branches mortared with mud and dried leaves, and when the platform is a few inches above the water, they construct a domed roof of mud and branches above it. Some lodges have one room, some more than one. The entrances to the lodges are narrow and steep, and all lodges have at least two (some have up to five). The most modest homes have one opening for entry and exit and another for food transport. The entire lodge, which may enclose a room more than 5 feet high, is plastered with mud before winter.

The kits are born in April, in litters from one to six. Their eyes open into the dark as soon as they are born, and within two weeks they begin to swim, accompanied by their mother. Beavers have a happy childhood. Posted around the edge of the lodge, they nose and push each other about, tumbling into the water. They race, wrestle, and dive in the pond, slapping their tails with abandon.

After a leisurely summer, the adults and the two-year-olds spend fall harvesting food for the winter, while the younger kits get in the way. Gnawing together, the beavers fell trees, slice them up, and drag them over their shoulders and under their arms to the pond, where the branches are cached in piles. Beavers store tons of wood in their pond, thinning the forest and removing saplings that were unlikely to survive in the over-story shade. Most animals grow thin in the winter, but beavers fare well and even grow fat when it's cold, feasting on inner bark and the roots and tubers of water plants, scrupulously digesting every bite twice and living amiably in their close quarters.

Since the kits stay at home through two winters, an estab-lished beaver colony includes the parents, this year's adolescents and this year's newborn kits—six to twelve beavers living together, cutting down trees, digging canals, and building dams up and down their valley. A beaver lives a dozen years or so, and works with its fellows for most of its life to build more and more dams, ponds, and lodges, until an entire valley becomes a mosaic of beaver handiwork. The original dam matures, and the old streambed runs like a ribbon through the marsh's center, almost filled with shifting silt and the billowing growth of bright marsh marigolds and water hyacinths. The trees have long since drowned, and old stumps break the water's surface, while willow sprouts grow thick along the wetland's borders. Over time, meadows emerge from the wetlands and thinned woodlands.

The beaver's dam is a telltale flag, however, and hunting beaver requires little more than persistence, which the Indians had plenty of. Trapping took place in winter, when beaver fur was prime. Since Indians did most of the trapping and traders wrote the journals, the stories may not be accurate in all respects, but according to these early accounts the Indians would block the stream above the dam with stakes, so that the beavers could not escape upstream, and tap the ice along the edge of ponds with chisels to sound out where the entrances to the burrows lay. At every entrance along the bank, they would make an opening in the ice, and then the lodge would be torn open. Those beavers that ran out of the lodge were clubbed, and those that ran into the mouth of a burrow in the bank were trapped there; the burrow was then broken into, or the beavers were pulled out with grappling hooks.

Beavers had also long been hunted with bow and arrow, and trapped by deadfalls and snares. The snare was made from rope, with a loop large enough to encircle an animal's head or leg. Sometimes a snare would be attached to a young tree, which would be bent over and held by a triggering device: when a beaver put its head or leg through the loop, the prop was dislodged and the tree sprang up, hauling the animal with it. The deadfall consisted of some large, heavy object—a boulder or a log—delicately positioned above the bait. The bait and trigger were smeared with castoreum, which beavers invariably investigate, and when a beaver went for the bait the deadfall came crashing down, pinning the beaver beneath it.

The first reference to a modern trap is in Leonard Mascall's 1590 British classic, *A Book of Fishing with Hook and Line . . . Another of Sundrie Engines and Trappes to take Polecates, Buzzards, Rats*. Widely deployed in Europe in the 1600s, the trap now known as a steel trap was pictured, and described as "a griping trappe made all of yrne, the lowest barre, and the ring or hoope with two clickets." The first steel traps used in North America were based on the Mascall trap, and had a round or oval baseplate. By the 1800s, the design had changed to a flat baseplate, with jaw pillars mounted at either end. The steel trap made simple work of harvesting beavers: instead of staking the stream and destroying the lodge, the trappers could drown the animals one by one. The traps cost $12 to $16 in the early 1800s, weighed 5 pounds, and were secured by a 5-foot chain with a swivel to prevent kinking. The trapper would wade up the stream to cover his tracks, and set the trap near the bank under 3 or 4 inches of water. To secure the trap, the chain was stretched to its full length and anchored to the streambed with a strong stake. A castoreum-coated twig was fixed above the trap, waving a few inches above the surface of the water. Any beaver that happened along would swim over to sniff the castoreum, place its foot on the trigger, and spring the trap's semicircular jaws. Diving down to conceal itself underwater, the beaver would find its movement restricted by the chain. If the trap caught only its paw, the beaver could gnaw the paw off and escape; if the upper part of the leg was trapped, the animal would try to gnaw through the chain—most often, it would be unsuccessful and drown. Even if the beaver succeeded in wrestling the stake out of the streambed, the combined weight of the trap, chain, and stake would eventually exhaust and drown the animal.

The beavers disappeared trap by trap, and hat by hat. But across the country, they disappeared by the tens of millions. When the beavers were removed, their old dams slowly collapsed, and the streams were released from the series of ponds and impoundments that had been built throughout the watershed. Each watershed lost wetlands, and the water that had once seeped quietly down to the aquifer now flowed to the sea, and flowed much more rapidly. Some of the springs and freshets that had bubbled throughout each watershed began to dwindle, while others disappeared entirely; in the undammed land, the water table soon dropped. Wetlands disappeared by the acre as the frontier rolled West.

Not only was there less water in the land but the water quality changed for the worse. In a land full of beaver, the stillness of ponds and wetlands had allowed sediment to settle, clearing the water and providing a large reserve of nutrients that stabilized the ecosystem. Over time, this collected sediment had formed rich bottomland valleys, building layers of topsoil. Without the dams and wetlands to slow the flow and allow the sediment to settle, the rivers became laden with silt.

Without wetlands, the runoff from the high flow of storms

and snowmelt was unimpeded, and storm or spring flooding could be two and three times higher than it was before. The swift-flowing water swept more soil into the stream—soil that was more likely to stay suspended—and muddy water blocked the sunlight from the algae. Without still warm ponds, the plankton were no longer as populous, and fewer minnows and insects found food. Without these tasty tidbits, the birds and animals that used to feast in the wetlands went hungry.

With fewer beaver ponds, there were fewer places for black ducks, ring-necked ducks, goldeneyes, and hooded mergansers to drop down to breed. Without the dams that maintained constant pond levels, muskrats and otters were either flooded or frozen out. Mink and raccoons, fond of eating the frogs, snakes, and suckers near ponds, found less food when the beavers were gone. The rabbits that had once nibbled safely among the brush and hidden among the felled logs were no longer so plentiful, and the red foxes found fewer to stalk. Moose and deer, which had browsed on the plants and waded in the cool water of the beaver colonies, lost their habitats. The beaver's wetlands had been home to a rich diversity of creatures of the air, land, and water, and without the beavers the fertility of vast areas was subtly reduced.

Today, the beaver has returned in part, but its numbers are nothing like what they once were, and we have forgotten that beaver wetlands once enlivened the now arid rangelands of the West. The total land area of the contiguous United States is 2.96 million square miles. Since the arrival of the Europeans, the beaver population of the United States has dropped from perhaps two hundred million to ten million. This decline in beaver population, and in beaver dams, caused the first major shift in the country's water cycle. If each of those pre-Columbian beavers had built only a single acre of wetlands, then an area of more than 300,000 square miles—a tenth of the total land area of the country—was once a beaver-built wetland. Now these wetlands are gone. The river of life receded when the water receded, and the primeval splendor of the land disappeared with the beaver's demise.

THREE

THE WOODS

The beaver and his dams were the first elements of the continent's natural water cycle to be disturbed. But as soon as settlement began in earnest, the forests came under assault as well. Greed certainly played a role, but the early reduction and runting of the American forests were driven mostly by cultural imperatives. The Indians saw the survival of broad stretches of forestland as tied up with their own survival. While they regularly burned back the underbrush—a practice we are just beginning to recognize as beneficial to forests as well as to Indians—the settlers brought with them a Eurocentric view of deep forest as an alien environment.

Wilderness was seen as mysterious and frightening; forested regions were thought to be unhealthy because the light rarely reached the ground, and the forest-dwelling Indians were seen as men "transformed into beasts" or as "bondslaves of Sathan." The English naturalist John Josselyn's 1672 description of the view north from a Connecticut mountain sums it up: "clothed with infinite thick woods," the landscape was "daunting terrible." Standing in aggregate, trees seemed to serve no useful purpose, and a pleasing vista was one that was cleared of trees, plowed, and planted. Moreover, the lumber could be put to good use.

Neither the Indians nor the Europeans understood that forestland is a necessary part of the natural water cycle. That understanding would await the publication in 1864 of *Man and Nature*, an erudite but readable work by a small-town Vermont lawyer named George Perkins Marsh, on how water, land, and forests work together. Citing resources in twelve languages (he spoke twenty), Marsh's rendering of the natural world and man's impact on land became an international classic. Using examples from across the Western world, Marsh showed that forests alter local climate, change the water cycle, and prevent erosion. His analysis of forests and watersheds is essentially correct today. Deforestation, said Marsh, changes everything. "The well-wooded and humid hills are turned to ridges of dry rock, which encumbers the low grounds and chokes the water-courses with its debris, and . . . the whole earth, unless rescued by human art from the physical degradation to which it tends, becomes an assemblage of bald mountains, or barren, turfless hill, and of swampy and malarious plains."

The changes, of course, were less dramatic than that, but real nonetheless. A forested watershed absorbs far more water than a treeless watershed, and where trees protect the soil relatively little silt is flushed into the waterways. Without forests, streams carry less organic matter and more silt. The silt clouds the water, reducing photosynthesis, and it infiltrates the gravel beds, which many fish use as nurseries. Without trees along the riparian zone, less organic matter moved into the waterways and less food fueled the stream ecosystem. There were fewer water striders skimming the still pools, fewer crawfish to eat them, less food to be found for the herons, and the wealth of life in the rivers dimmed.

It is estimated that about half the contiguous United States was once old-growth forest, with most of these primeval trees growing in the Eastern third of the country. Old-growth forests are characterized by fallen logs, snags, a multilayered forest canopy, and trees ranging in age from young to very old. The forest primeval of the New World was far from a dark tangle, though. Beavers had, over time, made forests into checkerboards of meadow and woods, and the Indians had for centuries managed the landscape on a massive scale with controlled burning.

In North America, fire was the dominant fact of forest history. The Indians torched large tracts around their villages, clearing the land for crops, stimulating a successional bloom of berries and browse for game, and, most important, refertilizing the soil. Burning the underbrush sweetens the soil pH and increases the rate at which nutrients are recycled into the soil. The improved production of fired pasturage has been evident for thousands of years, and the fertilizing qualities of ash are the basis of slash-and-burn farming (a form of agriculture that requires a lot of land and a low population density, for new tracts of land are required every three or four years when the soil wears out). Periodic burns allow grasses, shrubs, and non-woody plants to flourish, and conditions are improved for strawberries, blackberries, raspberries, and other gatherable foods. With no large domesticated animals to make manure, fire was a practical method of returning nutrients to the soil. The Indians fired cereal grasses annually, basket grasses and nuts every three years, brush perhaps every seven to ten years, large timber on a cycle of fifteen to thirty years or even longer. Fields, cleared by fire, were regularly burned both to enrich the soil and to keep the woods from encroaching.

Fire destroyed most tree and shrub seedlings, and thinned the forest canopy to create an open, parklike forest. Some tree species were highly resistant to fire, while others were either killed outright or so retarded and suppressed that the resistant species shaded them out. The fittest survived, often in pure stands. "The Salvages are accustomed to set fire of the Country in all places where they come, and to burn it twize a year, viz: at the Spring, and the fall of the leafe," observed Squire Morton, a 1632 visitor to Massachusetts. "The trees grow here and there as in our [English] parks, and make the country very beautiful." John Winthrop Jr., the eldest son of the first governor of the Massachusetts Bay Colony, noted in 1668 that in many parts of New England there were stands of only "a few large old timber trees of oak" 100 feet tall, with trunks 6 feet wide and a horizontal spread over 150 feet—groves of 400-year-old trees—while lithographs from an 1821 forest text show ancient stands of ash, beech, pine, and maple.

The Indians' regular fires increased the edge area between meadow and forest, and promoted open, single-species groves. These patches of varying habitat increased the diversity of forest types, creating ideal conditions for a host of wildlife. Edge areas supported more plants than did forests or grasslands alone, increasing the total herbivorous food supply and allowing the numbers of elk, deer, rabbits, porcupines, turkeys, quail, ruffed grouse, and so on to thrive. As these populations increased, so did the numbers of raptors, lynxes, foxes, and wolves.

The Indians not only maximized the available resources through regular burning but maximized their access to those resources by moving seasonally. The European explorers wrote of a land where fish were taken by lowering a bucket into the water, ducks could be clubbed from the sky, and grapes and strawberries garlanded the meadows; there may have been some literary license in all this, but there was some truth as well. Fish were easily caught during spawning season; strawberries were plentiful in June; huge flocks of ducks passed overhead in the autumn. The Indians planted their fields in spring, and caught fish that came up the rivers to spawn. During the summer, they moved into the woods to enjoy the abundance of squirrels, turkeys, berries, and other wild produce. At the end of summertime, the Indians returned to their villages to protect their maturing crops, gather the nuts from the forest floor, and prepare for their annual hunt. Harvest continued well into the fall, making it a time of feasting and plenty. But by the end of winter, food supplies often were low, and people grew thin in the last cold weeks before the fish swam upstream to spawn.

Colonists had a different standard of living in mind: as soon as they learned to cultivate corn (at first, mistaking the Virginia climate for that of the Mediterranean, they had planted olives and lemons along with rye, barley, wheat, and oats, and had nearly starved to death), permanent villages were built, fields were cleared, and pigs and cattle were turned loose to forage in the forests and marshes. Pigs had been forest dwellers before their domestication, and they multiplied rapidly in the wilderness. Soon hogs were considered "cattell for proffeit," and the trickle of colonists turned into a flood: in 1620, there were fewer than twenty-five hundred Europeans living on the Atlantic coastline, among roughly a hundred and fifty thousand Indians; by 1650 the number of colonists had swelled to fifty thousand, while Indian populations were vastly reduced by disease. And the Europeans just kept coming: two hundred and fifty thousand had settled by 1700; there were over a million in 1750, and nearly four million white settlers by 1790.

As the number of Europeans increased, so did their hogs, which rooted about in the woods, altering the habitats that supplied the Indians with animals for food and clothing. Hunting pressures from permanent settlements denuded the colonized land of game, and forests disappeared as pastures expanded. The net result was a simplification of the plant and animal communities.

The Narraganset sachem Miantonomo saw the future in 1642, just a few years after English colonists began to settle near his people's villages. "You know," he said, "our fathers had plenty of deer and skins, our plains were full of deer, as also our woods, and of turkies, and our coves full of fish and fowl. But these English having gotten our land, they with scythes cut down the grass, and their hogs spoil our clam banks, and we shall all be starved." And so they were, and the colonists prospered.

The change from Native American to English dominance in the New World saw a village system of fire-dependent shifting agriculture and hunting-gathering replaced by permanent farms, where crops and domesticated animals were raised within fences in pastures and fields. The Indians had depended upon the resources of a managed virgin forest, while the colonists wanted to clear it. Colonization and deforestation went hand in hand. Houses, barns, fences, barrels, wagons, and ships were all built of wood; wood was virtually the only fuel; and increasing quantities of it were consumed. In a little more than a century, the Eastern forests began to succumb to the axes and sawmills of the European settlers, and much of Massachusetts and Virginia was transformed into an ordered landscape of fields, barns, and clapboard houses.

The settlers of New England cleared the forest not for plantation agriculture but for timber and fuel. The Old World had felled its most accessible forests by the mid–1600s, and England welcomed a colonial wood source. As it turned out, shipping the bulky timber 3,000 miles across the Atlantic was expensive: it cost £8 to £10 per ship ton to transport timber from North America, while the freight rate from the Baltic forests was only nine to twelve shillings per ship ton. To develop a robust trade in timber, the New England colonists needed a market closer to home.

They found one in Barbados, which had been totally deforested by the slaves of British planters to make way for sugarcane plantations. New England supplied all the wood the sugarcane planters needed, for everything from buildings to barrel staves, and the market expanded as the West Indies prospered. By the time of the Revolutionary War, fewer than three hundred years after the arrival of the Europeans, Eastern America had changed from a climax forest harvested by many tribes of Indians to an agricultural society that annually exported millions of board feet of wood.

From 1789 to 1850, the total land holdings of the contiguous United States increased from 847,000 square miles to 2,975,000 square miles. Of this, 79 percent was originally owned by the federal government. In the 1800s, the government's policy was to transfer its land to states and settlers for development, and by 1900 two-thirds of the public land—more than half the land area of the nation—had been sold or given away. The preface to the census of 1810 sums up the federal attitude toward forested wilderness: "Our forests encumber a rich soil, an hundred or two hundred miles from the sea, and prevent its cultivation." To rid the land of its burden of trees, the authors of the census suggested that "we erect iron works, which require charcoal; of the maple trees we make sugar and cabinet wares; of the walnut and wild cherry, we make furniture and gun stocks; of the general woods we make potash and pearlash, of the oak, casks, and of the various trees, we make boards, joists, scantling, shingles, charcoal and ordinary fuel."

After 1815, the vast forests between the Mississippi Valley and the Appalachians were sold off, logging commenced, and most of the land was converted to farms. Ohio, for example, was nearly completely forested in 1800, when it was home to forty-five thousand pioneers. Eighty years later, there were over three million farmers, and 75 percent of the forests had been cleared. Indiana's lumber industry dates from the 1830s, when it specialized in oak and walnut, while Memphis, Tennessee, became one of the most renowned lumber centers in the United States, with over fifty sawmills.

In the nineteenth century, the American population doubled every twenty-three years. People were settling farther and farther west, and the demand for wood was enormous. Fireplaces provided the only heat for most of the population, and were astonishingly inefficient: between thirty and forty cords of wood per household is a figure that pops up again and again, with a single family spending about forty days a year chopping wood. By 1870, some five billion cords of wood had been consumed as fuel for fireplaces, stoves, industrial furnaces, steamboats, and railroads, and had accounted for the clearing of 200,000 square miles of forest. An additional 25,000 square miles of forest had provided the wood to build houses and barns, bridges and the wagons that rode on them, railroad trestles and crossties, ships and their barrels, fencing, furniture, bowls, and board sidewalks.

And the forests continued to fall for agriculture: cotton overtook tobacco as the premier cash crop in the early stages of the Industrial Revolution; by 1820, the United States had become the largest cotton producer in the world. Like tobacco cultivation, continuous cotton cultivation is wearing on the soil, prompting farmers to move on in search of fresh land. In the mid–1800s, the frontier of cotton growing spread steadily westward from the Eastern seaboard states of Virginia, the Carolinas, and Georgia into Alabama, Mississippi, Louisiana, and Texas. Between 1850 and 1860 alone, American farmers cleared 31,250 square miles of timberland.

In the previous century, financiers in New York, Boston, and Philadelphia had bought up the timber rights in Maine for 12½¢ an acre, and they cut as fast as they could. It took a hundred years or so to prove that the Maine woods were finite, and by the 1830s lumbermen needed new forests. The Erie Canal, completed in 1825, was soon carrying loggers westward by the thousands and floating millions of board feet of timber eastward. Albany, the canal's eastern terminus, became one of the busiest lumber markets in the world, and remained so for forty years.

The logging frontier cut through New York and Pennsylvania, and then paused in the enormous white-pine forests around the Great Lakes. In 1836 a tract of timberland on the St. Clair River in Michigan was sold to a Maine lumber baron, and one after another, the Eastern lumbermen moved in. Timberland in the Great Lakes region sold for $1.25 an acre, the standard fee for land in the public domain. Logging and the lumber trade boomed in the Lake states, making great fortunes for the timber barons. Trespassers made money too. The entrepreneurial cutting of trees on government land for private profit was especially common in the Lakes states. But even before these vast new forests reached peak production, it was perfectly obvious that here, as before, the prime timberlands were bound to dwindle. In the 1840s, the demand for lumber was increasing as fast as the burgeoning American population. And so the lumbermen moved farther west.

The monarchs of the Western forests dwarfed the trees that grew in the East and the Great Lakes region. The Eastern white pine, the tallest tree in the East, averaged 100 feet in height and rarely grew to more than 200 feet. The Western hemlock averaged 150 feet, with some trees growing to over 250 feet, while the redwoods and the Douglas fir both averaged more than 200 feet and sometimes exceeded 300 feet. (The tallest known conifer is a Douglas fir that tops 385 feet.) The coastal redwoods grew in a 500-mile-wide strip on the fog-shrouded coastline from the Oregon border south to Monterey Bay. Two-thousand-year-old trees weren't uncommon, and a substantial redwood could tower 350 feet, from a base 15 feet in diameter. The inland Sierra redwoods, or giant sequoias, caused a sensation when loggers first came upon them, in 1852. The sequoias ran to diameters of 15 and 20 feet; some of the grandest had trunks nearly 40 feet across. For commercial loggers, the redwoods and sequoias were a king-size bonanza. The wood was resilient to weathering and rot and made unsurpassed shingles, house siding, and railroad ties. The Sierra groves, which were never numerous to begin with, were cut with unflagging enthusiasm. Redwood trees filter the fog through their glossy needles, and drip so much water that a rain gauge on the ground below may record beteen 40 and 70 inches of precipitation a year, while a gauge at the same site clear-cut may record only 20 to 25 inches. Redwood seedlings need this moist microclimate to thrive; they dry up and die when the forest is cleared. By the time the wholesale destruction was checked in 1890, the sequoias were an endangered species. In the end, it was the towering Douglas fir, with an average diameter of 8 feet and a range from British Columbia to New Mexico, that became the greatest source of lumber the world had ever known.

The Pacific Coast's first commercial logging venture was launched in the Northwest in 1827, and by the mid–1840s the straight, strong timber of the Northwest forests was in demand throughout the Pacific basin, from South America to Australia and China. The Sitka spruce, pound for pound, is the strongest wood in the world: it makes the best ladders, and the warplanes of the First World War were built of it. The elegant, ginger-scented Port Orford cedar, which grew in commercial quantities only on a short stretch of the Oregon coast, took a fine finish and was in such demand for coffins in the Orient that the naturalist Donald Peattie wrote in 1950 that there was more of it underground in China than standing in Oregon. The Western hemlock became the chief source of wood pulp for the paper industry; the ponderosa pine—strong, fine-grained but rough textured—was good for almost anything; and the Douglas fir, straight-grained and tough, held nails, screws, and bolts more securely than oak and was therefore favored by home-builders and shipbuilders alike. The sugar pine, light and durable, didn't shrink or swell much in the weather, making it an excellent material for shingles, doors, and window sashes;

and since it imparted no flavor or scent once the sweet sap dried, sugar pine became the lumber of choice for fruit and vegetable crates and for packing up tea, sugar, and spices.

West Coast logging would remain an export business until a substantial local market developed for lumber. In 1847, the entire non-native population of the West Coast numbered no more than twenty-five thousand people, including the English settlements north of the Canadian border, the logging communities, and the Mexican towns and other foreign enclaves dotting the California coast. East of them was nothing but wilderness for 2,000 miles, all the way to Missouri. But at the beginning of 1848, gold was discovered in California, and the rush to mine it triggered the largest mass migration in United States history. Billions of board feet were cut for the houses and towns of the new settlers, and billions more for the crossties and trestles of the five railways that spanned the continent by 1893. Billions more board feet were exported to build railroads in Chile, Peru, and China, to shore up gold mines in Australia and diamond mines in South Africa, and to build the navies and merchant ships of European nations.

It took little more than a century for settlers to make a substantial difference in the extent of the American forestland. In 1790, when nearly four million people were confined to the Eastern seaboard, forests covered about a million square miles of the United States. By 1850, as settlements spread west, about 40 percent of these climax forests had been cleared. By 1870 the population was nearly forty million, and over 60 percent of the original forests were gone. It was a profligate era, and the natural resources seemed almost unlimited.

In the meantime, however, New England transcendentalism was presenting wilderness and forests in a new light. Transcendentalists believed in the essential unity of creation and beneficence of nature. Ralph Waldo Emerson and Henry David

Thoreau were its chief proponents, a pair of archetypical Americans who were among the first to speak for the wilderness. "In the woods, we return to reason and faith," and become "part and parcel of God" wrote Emerson in his first book, *Nature* (1836). Between them, Emerson and the author of *Walden, or Life in the Woods* (1854) set out philosophical guidelines for an unprecedented movement to preserve wild country, as the novels of James Fenimore Cooper and the landscape paintings of George Catlin, Thomas Cole, and George Inness gave literary and visual expression to the revolutionary notion that forests have value while standing. Wilderness began to be seen not as "daunting terrible" but as providing the freedom and solitude that human beings needed for their intellectual and spiritual regeneration.

With the ground prepared by Emerson and Thoreau, John Muir's writing in the 1870s articulated the environmental idea with a fervor and enthusiasm that commanded widespread attention. Major publications vied for his articles—on glaciation, on the wonder of the trees, on the miracle of nature, on the necessity of public ownership and government management of forestlands. Every book he wrote became at least a minor best-seller.

The son of a Scotsman so stern that the children's growth was stunted by spartan meals and dawn-to-dusk farm labor, Muir had a mechanical bent, and tinkered from an early age; he whittled clocks with Rube Goldberg functions, and devised tricky mousetraps. He was locally renowned as an inventor when he enrolled at the University of Wisconsin, where some of his contraptions—an alarm clock that tips up the bed and a study desk that changes your book every fifteen minutes—are still on display. As an adult, Muir developed a mystical bond with the mountains and would become physically ill when away from them for too long. He lived for ten years in Yosemite, and the greatest men of the day came to camp with him. He botanized, philosophized, and many accounts say that he looked like Christ when he came down from a week in the mountains. "Ink cannot tell the glow that lights me at this moment in turning to the mountains. I feel strong to leap Yosemite walls at a bound. . . . I will fuse in the spirit skies. I will touch naked God," he wrote, at a time when Victorian ladies were draping their pianos with shawls to cover the legs. Muir midwifed the revolution in environmental consciousness; the first land reserved for government ownership was Yellowstone National Park in 1872, and three national parks were added through Muir's advocacy in 1890—Yosemite, General Grant, and Sequoia, all of them chosen for their scenic value.

The new environmentalism was in good part responsible for the creation of the national forests. By the 1880s, the headlong process of land disposal had resulted in the transformation of hundreds of thousands of square miles of forest to stumplands and fields, and transcontinental railways allowed us to survey the damage for the first time. The first good forest maps were published in the 1880s, and the preface to that year's census reads, "The American people must learn that a forest, whatever its extent and resources, can be exhausted in a surprisingly short space of time through total disregard in its treatment." It had become obvious that without some form of forest management, the country would be facing a national shortage of wood.

At the same time, Americans were beginning to notice that the frontier was closing and the wildlife was disappearing. The beavers were gone. The passenger pigeon, which had numbered in the billions, was commercially extinct by the early 1880s and gone forever by 1914. The bison, which numbered sixty million or more when the colonists came, were cut down to one herd of a million in 1890; by 1892, there were eighty-five bison left in Yellowstone Park and a very few surviving bands elsewhere. The debates about preserving the forests on the public lands came to a head against a background of diminishing wilderness, real and imagined timber shortages, concern over widespread clear-cutting and timber trespass, unchecked mining and grazing on public lands, and the desire for some kind of fire control. In 1891, a provision was inserted into a land bill which allowed the president to reserve forest lands, and by 1910, thanks to President Theodore Roosevelt, nearly 250,000 square miles of forest had been reserved for government management—about a fifth of the standing timber in the country.

It's not just timberland that the government owns. Including Alaska and Hawaii, nearly 29 percent of the country is public land; of the contiguous United States, the federal government holds title to about 20 percent, or 626,000 square miles of land. The federal government's share of individual states ranges from less than 1 percent of New York and Kansas to 83 percent of Nevada, including almost half of all Western land. We have created an American commons. The ownership of land is not mentioned in the Constitution as a legitimate function of government, yet it seems that the system of public land is here to stay. By the turn of the century, it had become clear that private land ownership and capitalism spelled the rapid depletion of a region's resources; any red-blooded American pioneer would cut the last tree or kill the last buffalo for private gain. Public land ownership—the American commons—provided an ecological buffer against the depredations of capitalism.

The creation of national forests to ensure a long-term wood supply and protect the watersheds from uncontrolled lumbering, forest fires, mining, and grazing was the first conscious attempt to use our commons as a renewable resource. But in the early 1900s, most forests still lay in private hands in the Pacific Northwest, the South, and the Lake states. Three companies—Weyerhauser, the Northern Pacific Railway, and the Southern Pacific Railway—had accumulated 11 percent of the nation's standing timber through railroad, wagonroad, and swampland grants; the government held 20 percent, and the rest was split between farm woodlots and forests owned by individuals or small corporations.

The rate of deforestation slowed as the settlers moved onto the Great Plains, but by 1930 only 13 percent of the country's original forests were still in existence. Today, less than 3 percent—that is, roughly 6,250 square miles—might be said to remain in anything like its primeval state, and virtually none of this is in private hands; in total, it amounts to a single tract of land 60 miles long and 100 miles wide out of all this great land, and we're quibbling over how fast to cut it.

Forest canopies, especially old-growth forest canopies, are teeming with species. The rain-forest canopy was thought to be uniquely diverse until 1991, when the ecologist Meg Lowman built a walkway high between two big oaks near Williams College, in western Massachusetts. There, as in the rain forests of Panama and Costa Rica, Lowman's students found a world of weird aerial plants and unknown insects. Since then, hundreds of new species of insects unique to the North American canopy have been collected from the mats of moss and lichens that festoon the upper levels of old-growth forests. Bats abound, and flying squirrels and red tree voles spend their lives in the canopy. Some birds depend on it: as they hunt for flying squirrels, the spotted owls require the overhanging branches to protect them from being eaten by goshawks and horned owls. The marbled murrelets, also endangered, nest in the moss beds high above ground, and need the overhanging branches to protect their young from the jays that prey on them.

It is now thought that the lichens of an old-growth forest provide up to three-quarters of the nutrients that an enormous tree needs to survive. Lichens are organisms in which a fungus and an alga live together symbiotically—a relationship of two cooperative species creating a single organism which was first noted by Beatrix Potter, a self-trained botanist. Lichens and other epiphytes living high on old trees may weigh four times more than the foliage. Lichens feed on the nutrients that they filter from fog, rainfall, and dry floating particles in the air. A single old-growth tree may carry a thousand strains of a single species of fungus, while tree plantations have a very low diversity of these organisms. The vast filigrees of lichens that live on the trees collect and store many of the nutrients that would otherwise float by with the wind or wash away in runoff.

Many epiphytes live only in ancient forests, and some don't even appear for the first hundred and fifty years; the full range of lichen species may not be present until a forest is four hundred years old. Some strains of fungi defend their homes by producing compounds that kill tree-eating insects. Lichens live in slow motion, and can take fifty years to reach the size of a head of lettuce. Researchers suspect that lichens disperse slowly to new trees and must wait for specific conditions (like the crotch of a big, shaded lower limb) to establish themselves.

The point here is not "Save the Lichens." It is that we do not know how to grow big trees—or even how a big tree grows. Replacing an old-growth forest with a tree farm amounts to a total loss of habitat. If we continue whittling away at the last small fraction of old-growth forests, we will also lose the biodiversity necessary to bring them back.

When the national forests were set aside at the turn of the century, total forested area in the United States was slowly increasing, because trees will usually grow after a forest is felled if the land isn't farmed. By mid-century, farmland in New England had reverted to forest, as industrialization took hold and as farming began in the Midwest. After 1880, fields in the mid-Atlantic states started growing up in second-growth forest as the Plains states opened. Food, once provided by marginal farms in the Northeast and Southeast, was now grown on fertile prairie topsoil, and second-growth forests started creeping over the wheatfields of Vermont and the old tobacco and cotton plantations of the South. The adoption of the internal combustion engine decreased the amount of needed cropland by displacing workhorses, which ate quite a lot. (When the horse population peaked at some twenty-five million in 1900, about a quarter of all cropland was used to grow horse feed.) Workhorses virtually disappeared when the tractor and car took over, and fields that had been farmed for horsepower were freed up for food production. Land pressure was also reduced by improvements in farming techniques: the industrialization of our agricultural system reduced pressure to clear more forests by increasing the crop yield per acre. The total square miles of harvested crops is about the same today as it was at the turn of the century, but the yield per acre for most crops has more than doubled in that time.

Most important, fossil fuels supplanted wood as the fuel of choice. Beginning in the mid–1800s, coal was used to heat cities, and after the turn of the century the consumption of coal, oil, and gas spiraled upward, with corresponding decreases in wood consumption. We also use wood more efficiently now. Building a house using modern construction techniques requires about a third of the lumber used for an equivalent post-and-beam structure; and the bandsaw replaced the circular saw, reducing sawdust loss. Topwood is now slashed to rot back into the ground, or burned to add potash to the soil. Slabs are chipped for paper pulp and strandboard, and the bark is sold as garden mulch. Most of the hardwood topwood ends up in woodstoves, which put out eight or ten times as much heat as an open hearth. A 1903 report on forest management estimated that about 65 percent of a tree was wasted between its felling and its conversion into the final product; less than 10 percent is wasted today.

Forested land increased only haphazardly, however—until the 1930s, when the country was faced with an ecological crisis as well as an economic one. Hundreds of thousands of acres of gullied and eroded Southern farmland were abandoned as farmers fled West, and the federal government stepped in. The Civilian Conservation Corps, established in 1933, planted millions of trees to stop the erosion, rebuild the topsoil, and provide a plantation timber resource while encouraging adjacent private forestry. The Tennessee Valley Authority reforested watersheds as part of its mandate, and in 1934 the Forest Service embarked on the Shelterbelt Project. The shelterbelt was to be a 100-mile-wide strip of trees running from Canada to northern Texas along the line where precipitation drops to 18 inches—roughly, from Bismarck, North Dakota, to Amarillo, Texas. Millions of trees were planted in the Plains states, rehabilitating the land and increasing the nation's forestland. Today, we have 20 percent more forested land than we did at the turn of the century.

There are tremendous differences, however, between the forests of today and the primeval forests: one is that the trees are a whole lot smaller, not only because they're younger but also because the genetic stock of our forests has been debased by the practice of high-grading. Early logging was labor intensive and therefore selective: the largest trees went first, and they were larger than we imagine trees can be. A table from around 1800 gives the dimensions of trees in Vermont—measurements that "do not denote the greatest, which nature has produced of their particular species, but the greatest which are to be found in most of our towns."

	Diameter	Height
Pine	6 feet	250 feet
Maple	5¾ feet	From 100 to 200 feet
Buttonwood	5½ feet	From 100 to 200 feet
Elm	5½ feet	From 100 to 200 feet
Hemlock	4¾ feet	From 100 to 200 feet
Oak	4¾ feet	From 100 to 200 feet
Basswood	4¾ feet	From 100 to 200 feet
Ash	4¾ feet	From 100 to 200 feet
Birch	4¾ feet	From 100 to 200 feet

Today, the largest trees in these Vermont villages are rarely three feet in diameter, which is roughly half the girth of the largest trees two centuries ago. When a forest is high-graded, the largest trees are taken and the runts are left as seed stock. The trees still standing come in different sizes, of course, and for the next cut the logger selects the largest of the runts, leaving the smallest runts to produce more runts—the status of much of our forested land today.

Another difference between our forests and those of yester-year is a consequence of our belated understanding of the role of fire in a forest ecosystem; many of our forests bear the brunt of well-intentioned mismanagement. For eighty years, the U.S. Forest Service did its best to squelch all wildfires, and the result in many areas is a sickly forest of subdominant species. The Blue Mountains of Oregon are a classic example: Once, they were covered with towering groves of fire- and insect-resistant ponderosa pine and Western larch; the low-intensity fires that regularly crept through the forest eliminated brush and other tree species. The ponderosa pine was logged off, fires were aggressively quenched, and the trees left behind were the lodgepole pine and the white fir, both of which are vulnerable to insect infestations. Today, the forests of the Blue Mountains are dying of insects and disease, with deadwood building up on the forest floor and the chances of a catastrophic forest fire increasing every year. Likewise, in forests across the nation selective cutting has resulted in smaller trees and a skewed species balance, while fire suppression has reduced natural culling and reduced concentrations of fire-resistant species, resulting in what the University of Vermont forest ecologist Hubert Vogelmann characterizes as "a forest of sticks."

Almost gone are the ancient groves of ash and maple that the Indians made, the towering stands of sequoias and redwoods, and the wonder of a world of 8-foot-diameter Douglas firs. Forests are seen as a crop, and the trees today are shadows of what they once were. The reduction of forested land from the time the Europeans landed until today amounts to about half. State- and government-managed forestlands total 206,250 square miles, while privately owned forestland covers over 600,000 square miles of the United States—which is to say that this country's waterways have suffered the equivalent of a kidney removal.

~

FOUR

THE VOYAGE OF
RAINFALL

From the water's point of view, a forest is a forest whatever its age. When rain falls on a forested watershed, the canopy breaks the force of the falling drops, and a resilient mat of twigs, leaves, moss, and dead and decaying plants keeps the soil from splashing. Over half the rain that falls on a forest goes directly back into the air through evaporation and transpiration, creating a moist and clouded microclimate. Some of the rain slowly percolates down through the forest soil to become the groundwater that reemerges in springs, and some of the rain runs over the surface of the soil and enters the waterways. As the leaves in a tree connect to a stem, twig, branch, limb, and on down to a main trunk, so the runoff collects into rivulets that flow to a brook joined by another and another, until ultimately a great river reaches the sea.

Rain that falls to the earth is isolated into different watersheds (or river basins, or catchments) by folds of the earth's lithosphere. Water erodes the surface, and runoff always carries silt along with it; these tiny rock particles, less than ¼₀₀ inch in diameter, are fine enough to become suspended in the water as it makes its way downhill and coalesces into larger flows. Along the way, organic matter and silt are retained behind natural dams or filters formed by geological features, by trees and branches in the streambed, and by beaver dams. Life in the waterways is designed with these inputs in mind: the silt provides atoms of inorganic elements, while the plant debris provides streams with organic matter, or food. The input of debris and silt to a stream is predictable: somewhere between 65 and 82 percent of the organic debris that falls into small streams every year stays in the streambed; the rest is flushed downstream.

In the headwaters, water usually flows in shallow, turbulent, boulder-strewn streambeds. If the riparian zone is treed, the water is shaded, running, and cool. Plant plankton are slow to multiply in those conditions, and contribute few calories to the food web. Mosses, growing on boulders, bedrock, and wood, cover more than 20 percent of the stream surface, and there is a patchy blue-green algal community intimately associated with the mosses, along with sparse populations of diatoms. Most of the nutrients that feed the stream ecosystem are in the leaves, twigs, limbs, and trunks of the trees that tumble into the water from the riparian zone. Tree trunks in the stream are slowly broken down by physical abrasion against boulders and the streambed, and also by bacteria, fungi, wood-gouging beetle larvae, leaf-shredding stoneflies, and snails. Leaves are processed relatively quickly, and the organic debris continuously breaks down into smaller particles that are available for microbial consumption.

Nutrients cycle in still water, rising from the bottom sediments to the top of the water column and sinking again. In running water, nutrients are swept along in the current, so instead of cycling in place they spiral down the river system. In a stream flowing through an old-growth forest, logs obstruct the channel, creating two types of habitat: the wood itself, and the pools where organic debris is deposited. The little dams formed by fallen trees, each of which retains organic matter and nutrients, tighten the nutrient spirals and increase the productivity of the stream ecosystem.

Water that flows in a river moves like a corkscrew, twisting in on itself. The water on the river bottom is slowed by friction as it moves over obstructions on the riverbed, while the water on the surface of the river flows more quickly. Where the river bends, the faster flow at the surface pushes against the bend's outer bank and erodes it, while the slower, siltier water at the bottom slips to the inner bank and drops some of its load of sand or gravel. This process creates the meanders found on lowland rivers.

Within a stream channel, the variable flow carves a pool at the outside edge of each curve and a bar of silt at the inner edge, with riffles of gravel on the straightaways. Pools and riffles are distinct habitats, with characteristic communities. Riffle dwellers are adapted to swifter, shallower water. Some are small and hide between the rocks; others hold onto the substrate. Many fish lay their eggs in the protected, aerated shelter of the gravel beds. Deeper pools are able to support not just the larger animals that range throughout the water column but also the bottom dwellers, which process the organic matter that settles to the bed.

Rivers in open country support more complex communities than woodland streams, as a rule. With more sunlight, there's more algae, supporting more zooplankton and providing a broader base for the food chain; river water supports a greater combined weight of biomass—living tissue—including plants, insects, amphibians, fish, flesh, and fowl.

The water in a river is always moving, and the bends in the river are always narrowing. As the point bars and cut banks become more pronounced by the twin processes of deposition and erosion, a meander becomes more extreme. The river will eventually cut through the narrow neck of land and create a new channel, leaving an oxbow lake behind. That small lake, over time, will fill with sediment and revert to floodplain—which will be carved again in another way, by a new channel, many ages later, as the earth's surface and climate shifts. Over a period of several hundred years, a meandering river channel writhes like a snake, with the meander loops sliding downstream, throwing off oxbow lakes as they go. The local features of a stream or river change and eventually disappear, but on the geologic timescale the overall pattern between the river and the land remains the same.

The dynamic equilibrium between the waterways and the land create a corresponding dynamic equilibrium of life within a river system. Successive plant and animal communities occupy a meander loop as it is transformed from an active channel to an isolated oxbow intermittently connected to the main flow during floods, and finally to a wet depression on the floodplain. As long as the river system keeps creating new cutoffs, a succession of habitats suited to each type of community is maintained, and all stages will occur on the floodplain. If the channel is "stabilized" and the floodplain leveed, the organisms that depended on sandbars, undercut banks, oxbows, and floodplains begin to disappear.

Although upland streams are typically steeper than lowland rivers, the water moves more slowly in the smallest upland streams than it does in a silt-bedded lowland river. The water arcs and curves in the smaller channel, incorporating oxygen into its flow, leaping and churning against boulders and logs. As the narrow upland streams converge into larger flows, water that had been expending much of its forward velocity swirling about now shifts to a laminar flow—straight ahead—as the streambed changes to cobbles, gravel, and sand. Eventually the flow becomes a deep, meandering, silt-bedded lowland river, establishing a delta (where much of the silt load is dropped), interlaced with estuaries, where it meets the sea. The deltas of the largest rivers ultimately include many thousands of square miles: the Mississippi River has built and abandoned six enormous deltaic lobes in an orderly cycle over the last eight thousand years.

Winter precipitation melts in a few spring weeks, and rainstorms deposit large volumes of water in a short period of time, so rivers flood regularly, donating sediment and organic matter to the floodplain. The channel of a river is only a fraction of the total area annually flooded; most of the aquatic productivity in a large river-floodplain ecosystem is in the floodplain rather than the river channel itself. The flood pulse of rivers fed by forested watersheds is timed predictably and maintains a well-established floodplain. During the late summer and fall, when the flow in forest streams is scanty or intermittent, deadwood litters the streambed. But when the water runs high in the spring, formerly insignificant flows can move many tons of sediment and driftwood. The stream channel is reworked during the high flows of spring, and the silt that accumulated in the interstices of the gravel beds is flushed out, renewing the sites of fish hatcheries.

Trees grow along the banks of streams, and they continually slip into the waterways. Stream banks cave in, windstorms blow trees over, and snow and ice pull trees down, tearing off their limbs and branches. With logs in the way, the stream flow is diverted into new pathways, increasing the complexity of the channel. Much of the driftwood that enters the waterways is swept in from higher land in debris torrents triggered by avalanches or mudslides. In the mountainous old-growth watersheds of Alaska, I've seen 15-foot-high haystacks of trees collect where small tributaries enter an intermediate-size stream. One morning after a recent storm dropped 8 inches of rain near my farm in central Vermont, it looked as though the whole forest had been in motion. Where the rampant flow undercut a stream bank, I saw a pine fall down and knock another over. Branches had broken in the high winds; the stream's banks had shifted noticeably, and moss-covered logs in the channel had been deposited a few yards downstream. Forests move more than we realize, and trees are constantly entering the waterways.

When a tree trunk tumbles into a river, it sculpts the channel on a larger scale than most riverborne debris will. Some logs form isolated snags, which may eventually promote the formation of sandbars that become small islands; other logs form logjams that are washed out to sea seasons later. The enormous quantity of big wood in the coastal rivers that once drained old-growth watersheds is reflected in the records of the U.S. Army Corps of Engineers, an organization charged with keeping the nation's waterways navigable. Between 1890 and 1917, 34,827 drifted trees were taken out of a total of 74 river miles in Coos and Tillamook counties in Oregon—along with 811 scowloads of wood and 1,751 boulders. All in all, along those miles of the Coos, Tillamook, and Coquille Rivers, the Corps removed about 470 trees per mile over twenty-seven years, or a tree for every 11 feet. Big driftwood shaped the river channels, and it covered the shore wherever a river met the sea.

Most of all, the land fed the sea great logs, and these drifting trees would occasionally mat together into floating islands. The Atchafalaya, a river that once built one of the lobes of the Mississippi's great delta, produced in the last decades of the 1700s a raft of logs 10 miles long, upon which grew trees that were 60 feet high by 1835, when it was cleared away. In the nineteenth century, floating islands of matted trees were sometimes seen in the Indian Ocean 100 miles from the mouth of the Ganges.

In addition to creating islands in the sea, every log is an island unto itself. A floating tree is a secure point of attachment in a drifting world, and provides a steady base of nutrients that supports complex communities. Floating debris becomes a focal point for life in the desert of the open ocean, and such communities can be seen from great distances. Algae and other aquatic plants attach themselves to a tree trunk; burrowing larvae rasp away at the wood; barnacles catch hold, while small fish eat the algae and enjoy the shade as well. Over time, larger fish arrive to eat the smaller fish, and birds circle above to dine on the swimmers. Some birds will alight on the log to rest their wings, and will leave behind a pile of phosphorus-rich guano filled with seeds. Sometimes a mangrove seed will sprout, and eventually, since the mangrove is a salt-resistant species, one of these sprouts may grow into a mangrove tree rooted on the log. In a number of years, a tree that once stood as a member of a forest community can create another world, supporting a column of life that extends from large fish swimming deep underwater to birds flying high above, connected by the isolated fragment of new land in between.

Islands, which range in size from Australia to a floating log, are hotbeds of evolution. When an ecosystem is isolated, the limited number of species that arrive may evolve quickly in the new environment, dividing up the resources between them. On Mauritius, without predators, the doves became 50-pound dodos; on the Galapagos Islands (named for their tortoises, from the Spanish *galapagar*), fourteen subspecies of tortoises once did much of the grazing; on Sulawesi, the macaque monkey diversified into seven species. If islands are places where evolution is at its peak, these seaborne logs must have contributed to the genetic diversity of the oceans themselves. Reasonable extrapolation from the Army Corps of Engineers' numbers shows that North American rivers could have fed the sea a few million logs every decade. If these logs averaged only 10 feet long, they would have totalled thousands of miles of floating edge between the land and the sea. Until quite recently, floating trees would have increased the ocean's edge—and productivity—on a global scale.

$\sim$

When trees are standing in a forest, they act like giant water pumps, directing water back to the atmosphere. The roots are connected to the leaves by a simple plumbing system: the sun, through the leaves, feeds the roots, and the roots collect water and nutrient elements for the leaves, moving liquids up and down in the sticky cambium layer that usually lies right below the bark. Water and nutrient elements move up the tree in the xylem, a layer of tubes inside the cambium sheath. The food from the leaves travels down to the roots through the cells in the outer layer of cambium, the phloem. The cambium layer grows fastest in the rainy season, and the cells are large and light-colored. In the summer and fall, growth slows and the cells become thick-walled, smaller, and darker, creating cross-sectional annual rings. Old phloem dies and becomes bark; xylem becomes impregnated with lignin as it ages, which binds together the water-carrying tubes into the tough, fibrous substance we know as wood. The trunk never gets taller; it only grows wider. All upward growth takes place in the tips of the branches, so if a nail is driven into a tree at eye level, years later the nail will be at the same height though the tree will be taller and the girth will have expanded. Wood holds the tree up, but the cambium layer does all the biological work, so a hollow tree can be just as healthy as a solid one (though it's more likely to tip over).

Like the limbs above ground, roots branch and rebranch. Tapered, they are in effect wedges driven into the ground, breaking up the subsoil and retrieving minerals from the parent rock. Water and nutrient elements enter the tree through tiny new hairs behind the root tip, which probes and burrows into the soil like a curious finger. (Darwin described a root tip as behaving "like the brain of one of the lower animals.") Each delicate root hair wraps around a grain of soil and passes moisture and dissolved minerals into the cambium sheath. Water and trace elements are channeled up to the leaves, where sunlight, water, and a few stray elements are converted into sugar and proteins that move down to feed the growing roots. Root hairs live only a few days, so the roots are constantly growing, and over time the root hairs entangle so many grains of earth that the soil is held firmly in place.

As the leaves rely on the root hairs, so the root hairs rely on mycorrhizal fungi for nutrient uptake—a relationship that has persisted since the earliest plants took root some four hundred million years ago. Mushroom roots, or mycorrhizas, are mantles of fine root hairs that many species of mushrooms form on plant roots. Some mycorrhizas live outside the host roots, others within; some live on many different species of plant roots, others depend on a single host species. As the mice, voles, and chipmunks tunnel through the forest floor, their feces inoculate the soil with mycorrhizal fungi spores. The mycorrhizal fungi collect water laden with nutrients leached from the topsoil and minerals extracted from the subsoil, and pass it on to the root hairs. Recent studies show that the fungi appear to collect nutrients selectively, as if the tree were somehow communicating to the fungi which elements were needed.

The tree recycles these nutrient elements as leaf litter, which improves the fertility of the soil. By the time a tree is full grown, the underground root system is enormous; a large oak, for example, has literally hundreds of miles of roots on an endless quest for nutrient elements and water. Each drop of water collected is transferred up the trunk to the leaves, and a mature oak may transpire 170 quarts of water a day.

Beneath the duff covering the forest floor, the soil is honeycombed with channels made not only by roots but also by burrowing animals, earthworms, insect larvae, and fungi. Between 50 and 60 percent of the total volume of healthy forest soil is space. Some of the passageways are less than $\frac{1}{100}$ inch across, holding water by capillary action. Larger passageways—those used daily by earthworms and insects—are connected to conduits a quarter inch or more in diameter, down which soil water moves rapidly. Rain that trickles from the leaves to the

forest floor fills the larger channels first, and then filters verti-
cally and horizontally from one channel to another, as it seeps
into the interstices of the soil. Much of the water that percolates downward to become groundwater will eventually
emerge at lower altitudes, as bubbling springs and seepage that
maintain the flow of creeks and streams. The smallest of these
streams are fed by rainwater that leaches through the organic
debris of the forest floor.

A single water molecule making its way through a stream-
and-forest ecosystem is on a biological Ferris wheel. A raindrop
may hit a leaf, trickle down to the bark of a branch, evaporate
to come down again as rain that flows into soil and is sucked up
by a root hair and transpired from a leaf—to become part of yet
another raindrop that comes down in a storm and runs over-
ground to a stream. Snowflakes that fall to the forest floor pile
up in a blanket tucked around the trees, shaded from the sun. A
layer of snow is made up of millions of hexagonal crystals of
such delicacy that their surface area is simply enormous. As long
as the air temperature stays below freezing and the snow has no
crust, the difference in vapor pressure between the air and the
surface of the snow crystal encourages water molecules to jump
ship: snow turns directly into water vapor without becoming a
liquid, in a process called sublimation, which is to say that snow
on the branches of a forest often sublimes. Snow may last for
months in a forest, slowly subliming and melting, going back to
the air and seeping down through the duff, which soaks water
up like a sponge. The snowmelt slowly trickles down to the
water table, and when every interstice is full and the forest soil
brims with water, the sluggish underground river moves down
the watershed, pushed along by the melting snow from above.

Less snow accumulates on forested lands than on treeless
ground, and it lingers from one to five weeks longer in spring
than the snow on a nearby field. The snow on a forested watershed melts over a longer period of time, reducing the high flow
of streams in the spring. Some areas of the Rockies receive 90

percent of their precipitation as snow, so deforestation there can lead to significant flooding. This, in turn, brings water shortages, for when water from rain or snow isn't stored in the land, the water table will drop.

Without trees, there are fewer leaves and less total surface area for evapotranspiration, so there's less moisture in the air. Without a protective canopy of leaves, the soil is struck with the full force of the storm. Individual raindrops are like little bombs, gouging, beating, and battering the soil, lifting and splashing it back and forth, churning it into a pasty mud that clogs the pores and passages in the soil. There is less biological activity in the soil in a field than there is in forest soil, so the earth is less tunneled and less water percolates down into the groundwater. As the groundwater recedes, the springs falter, the streamflow slows, and some streambeds will eventually run dry.

When rain falls on a deforested watershed, the runoff races downhill in millions of little currents, combining into torrents of water that gouge and gully the hillside. Without trunks in the streambed to slow and deflect the flow, the stream becomes wider and shallower, shifting to steeper slopes and eroding the banks as it goes. The runoff sweeps along tons of topsoil, gravel, and stones, and dumps the spoils into the waterways. As the silt is moved into the water, the formerly narrow forest stream eventually splits into a broad swath of braided channels that alternately scour the landscape and go dry, shifting across a broad waste of gravel and debris.

In the final analysis, the most significant change wrought by deforestation may be that waterways no longer feed many trees to the sea. Streams that drain deforested watersheds don't carry much wood, nor do streams draining logged watersheds. Huge man-made dams remove logs from the waterways as well. Those floating columns of life—the logs that once enlivened the landscape of the open oceans and increased the productivity of forest streams—are disappearing.

FIVE

A SEA OF GRASS

Grasslands, the geographer Carl O. Sauer first observed in 1950, are the consequence not of climate alone but also of a complex set of periodic disturbances, such as fires and grazing. Grasslands develop in areas where occasional droughts dry out the vegetation, but they are not exclusively determined by rainfall. They grow where level to gently rolling land facilitates the spread of regular fires over large areas and accommodates large herds of grazing animals, which trim the vegetation. In the grasslands of North America, periodic droughts, high temperatures, and strong winds provided an ideal environment for the ignition and spread of fire, while great herds of buffalo and elk did the grazing.

The grasslands of the interior ran north/south in three vast irregular belts. The tallgrass prairie—the easternmost belt—was characterized by giant grasses that often reached a height of 12 feet. The tallgrass prairie got 35 inches or more of rain a year; it included most of Indiana and Illinois, and extended about 100 miles west of the western borders of Minnesota, Iowa, and Missouri. With less rain came the mixed-grass prairie of the Dakotas, Nebraska, Kansas, Oklahoma, and central Texas, where tallgrasses grew on the lower, moister land and shorter grasses grew on the uplands. The short-grass prairie, where short blue grama and buffalo grasses predominated, began where the annual rainfall drops to 20 inches or less and reached to the Rocky Mountains. It included northeastern Montana; the eastern stretches of Wyoming, Colorado, and New Mexico; and the Texas-Oklahoma Panhandle.

Humans and lightning subjected the Plains to repeated and extensive burnings for thousands of years, and the plant and animal species adapted to regular burning. Fire released the nutrients bound in the litter, which would otherwise be unavailable for plant uptake until the litter decayed. Early spring burning increased the number of shoots and consequent flowering stems of the grasses, by fertilizing the plants at the start of the growing season. Grasses produce biomass more quickly than the existing biomass can decompose, and if grasslands aren't burned, growth and flowering are suppressed and plant vigor visibly declines. Fire also encouraged the growth of grasses at the expense of woody plants; shrubs and tree seedlings were often killed outright, and woody plants that were top-killed and resprouted would likely be eaten by browsing elk or deer.

Tallgrass prairie reverts to scrub woodland in a matter of decades without regular fires, so these grasslands in particular are thought to have been the product of deliberate, routine firings by the North American Indians. The Indians used broadcast fire for pasturage, for reduction of brush and ease of travel, for slash-and-burn agriculture, for hunting, for warfare, and more subtly too: most evidence suggests that Indians frequently started grassland fires in order to modify habitat and attract wild game to the tender new growth. Grasslands were typically burned annually, with occasional unplanned holocausts from lightning and escaped campfires and periodic conflagration in times of drought. Fire was predominantly local, but the local burnings had enormous cumulative effects.

The anthropogenic grasslands of pre-Columbian North America extended virtually unbroken from east of the Mississippi to the Rockies. As they radiated east and west from these borders, their continuity broke down and individual grassy "barrens" appeared, including the Central Valley of California, Oregon's Willamette Valley, the Shenandoah Valley, and the celebrated Barrens of Kentucky.

The colonists had a language problem, for the English word "meadow" didn't describe the Indians' firebuilt grasslands. The terms "barrens," "openings," and "deserts" were often used, with the scholarly types resorting to "champion fields," from the French *campaigne*. The Great American Desert is from the French *déserter*, "to abandon," since the land was seen as cleared and deserted rather than as naturally desiccated, while "barrens" reflected the English perception of land without forests as infertile.

The Barrens of Kentucky, the eastern outpost of the tallgrass prairies, was a crescent-shaped meadow of 6,000 square miles surrounded by virgin forest. It was burned annually by the Indians and supported thousands of buffalo. John Filson, the first chronicler of Kentucky, wrote in 1784, "The amazing herds of buffalo which resort thither, by their size and numbers, fill the traveler with amazement and terror." As soon as the settlers moved in and the Indians were moved out, regular burnings stopped, the great herds were annihilated, and the land was quickly treed over.

After the explorers breached the Appalachians and the individual barrens began to give way to a vast sea of grass, the French explorers dubbed the ecosystem a prairie, from Latin *pratum*, for "meadow." The Spanish, meanwhile, called it savannah, but the settlers of the Great Plains came largely from the north, and the name that stuck was "prairie." "Grasslands" is a twentieth-century term.

In forested land, animal species have adapted to life in a variety of habitats, from below ground to the top of the canopy. In grasslands, there is limited vertical variation, and the habitats differ horizontally, from patch to patch, where different disturbances have left their mark. The Plains vegetation is not a standard mix of grasses and forbs (that is, leafy plants) but varies widely from place to place. There are about two hundred and fifty species of plants on the prairie: some, like the little bluestem, are generalists that live across the entire grasslands; others require specific conditions in order to thrive. The plant species can be roughly divided into tall grasses, short grasses, tall forbs, and short forbs. These plants live together in an ecological community, and have different responses to the stresses of prairie life. A nearsighted view of a patch of prairie shows an area dominated by a few common grasses, with roots that create a dense underground web, and a broad range of interstitial species.

Drought, grazing, and fire are the large-scale disturbances affecting the plant mix, creating a patchwork of slightly varying vegetation. Drought affects tall grasses more seriously than it does short grasses, so the dominant grass species will shift in response to prolonged local droughts. Buffalo and cattle much prefer grasses to forbs, so grazing increases the available space for interstitial plant species. Fire, which burns some areas and not others, appears to increase the dominance of some grasses, killing many interstitial species and the year's seed crop for the few species that are flowering when the land is burned—though many grasses can multiply even when their seeds are burned, via shoots that grow laterally below ground. Grasses protect themselves from drought by dying down to their underground organs. Their growing points lie below the surface of the soil, which also protects them from fire and overgrazing. Unlike a leaf, a blade of grass continues to grow from the base after the tip is removed, so grazed grasses produce leaf tissue throughout the growing season. Small-scale disturbances, like buffalo wallows, gopher mounds, prairie dog towns, and badger holes, provide a series of radically different habitats in this superficially uniform landscape.

On the Great Plains, most of the rain falls in brief violent storms in late spring and early summer, while the area from Utah north receives its precipitation in the winter and early spring. Most of the rainfall on the prairie clings to the vegetation, where it either evaporates or drips down to the soil, which is shielded from the direct impact of the rain. Some of the water will course over the surface as runoff, but generally the vegetation impedes its progress. A raindrop on the ground can sink into the soil because of gravity, capillary action, and air pressure (spaces in the soil are filled with air too, but barometric pressure above ground changes faster than the pressure below ground), but these are small forces; the difference in vapor pressure between water and air usually causes the water molecules to evaporate instead of entering the soil. If any raindrops do make it underground, the dense root systems take them up; in a semiarid climate, rainfall rarely percolates below the root zone to replenish the groundwater.

The prairie evolved with fire, and also with large herds of herbivores, which grazed on the grasses (and on the forbs when they had to). There once were perhaps sixty million buffalo on the grasslands, but the buffalo's gregarious nature didn't exclude other grazers. When Meriwether Lewis crossed the Plains in the spring of 1805, he described them as covered with "immence herds of Buffaloe, Elk, deer, & Antelopes feeding in one common and boundless pasture." Lewis and Clark reported killing over a hundred large game animals, including twenty-nine elk, twenty-eight deer, seventeen buffalo, fifteen black and grizzly bears, a few goats and bighorn sheep, and a "lion." The goats and the lion were what we today call mountain goats and mountain lions; species so closely associated with mountainous terrain that it has become part of their common name, and which have actually been sequestered there by hunters. Although mountain goats are now pursued in the highest crags, two centuries ago they roamed the Western plains. Similarly, elk now summer in high mountain meadows and come down to the foothills for winter, but in 1805 they were grazing on the prairie.

The herds were predominantly buffalo, though—"buffalo" is the common name for *Bison bison*, a branch of the Bovidae family that is not directly related to the African Cape buffalo or the Asian water buffalo—and they grazed the prairie cyclically, in groups as large as a few million individuals, feeding in one area for a week or two and then moving on when the pasture began to thin. They are huge animals, especially compared to the scrawny cattle of the time. A mature bull weighs over a ton and stands about 6½ feet at its hump. Most of its weight is up front, in the massive low head and powerful shoulder and neck muscles, which are built to sweep away snow from the forage. An unusually long spine dwindles bowlike to the haunches. The coat is blizzard-proof: a shock of black hair grows over the head like a hood, while curly brown fur covers the rest of the body. The hair on the animal's forepart is permanent, but the hindquarters shed annually, beginning in March. By early summer, the buffalo's hind end is nearly naked, and very attractive to bloodsucking bugs.

To protect themselves from insect bites, the buffalo would dig shallow ponds with their sharp, cloven hoofs. The wallows averaged about 20 feet across and 2 feet deep—large enough to allow the buffalo to cover their hindquarters with mud, which dried in the sun and formed a protective coating that lasted for several days. In drier areas, the buffalo would take dust baths in dry wallows, which were about 10 feet in diameter and less than a foot deep. The wallows dotted the plains wherever the buffalo roamed, providing drinking holes for animals and patches of moister ground that could support distinctive plant communities.

Buffalo wallows were dug in low areas, and they collected runoff as well as rainfall. The fate of the water that pooled in those little ponds was very different from that of the water that fell on the grasses. Only the uppermost layer of water in a wallow could evaporate; the rest seeped down to the water table, creating a zone of saturated soil that extended from the bottom of the wallow to the groundwater. Every wallow was a pathway for runoff and rainfall to percolate down to the water table. Civil engineers dig recharge ponds that look just like buffalo wallows, in order to increase the rate of groundwater recharge; the only problem is that a layer of silt will collect on the bottom of the pond and, over time, clog it. The buffalo's hooves prevented such a layer from forming, so an active buffalo wallow could be described as a perfectly designed groundwater recharge pond.

Given the lack of trees and the regular fires, many of the smaller animals on the prairies lived below ground. Prairie dogs, like beavers, are a keystone species—that is, one that significantly alters the ecosystem and provides habitat for auxiliary species. No one knows for sure how many there used to be, but it is thought that billions of prairie dogs once burrowed in the Great Plains. The black-tailed prairie dog (*Cynomys ludovicianus*) ranged throughout the mixed- and short-grass prairies; the white-tailed prairie dog (*Cynomys leucurus*) lived generally to the west, at higher elevations in the short-grass prairies. Two related species with white tails are the Utah prairie dog (*C. parvidens*) of southwest Utah and the Gunnison prairie dog (*C. gunnisoni*) of high short-grass prairies in Utah, Colorado, New Mexico, and Arizona. Prairie dog towns extended for thousands of square miles, with about fifty holes per acre. Indeed, their social structure and burrowing skills are so refined that prairie dogs might have taken over the grasslands but for *Pasteurella pestis*, the plague bacillus (also harbored in cottontails, marmots, mice, and a few other rodents); large-scale outbreaks have long been noted among them.

Prairie dog towns are underground mazes of 5-inch-diameter tunnels, which range in length from less than 20 to over 80 feet. The tunnels usually lie within the root zone, which may reach as deep as 10 or 12 feet in some parts of the Plains. The burrows have pockets, turn-around rooms, and nest chambers lined with grass. Some burrows have one opening, some burrows have several; some burrows have underground connections, others don't. In the process of constructing their towns of tunnels, the prairie dogs once moved tons of subsoil above ground, where they mixed it with topsoil and organic matter, including clipped grass and roots, feces and urine, insect parts, and other by-products of life; they made loam in a sandy world. The greens between the holes were kept clipped to expose predators, creating a short-grass plant community. New growth was tender and higher in protein than the surrounding grasses, so buffalo and cattle grazed preferentially on prairie dog towns. Since the grass around their towns was kept short, fire often passed by colonies entirely, providing an oasis of unburned vegetation where the rabbits, mice, and other small creatures could eat until the prairie greened again.

Life underground is more temperate than life on the Plains, and the burrows stay cool in summer and warm in winter. (In a 1954 study, Maxwell Wilcomb, of the University of Oklahoma, found that in the grasslands of that state the average temperature 4 feet underground ranged from 50°F in winter to 80°F in summer, while the surface soil temperature varied from −10°F to 120°F.) Snakes, cottontail rabbits, skunks, mice, box turtles, and burrowing owls live in abandoned prairie dog holes. Toads, lizards, and tiger salamanders use the burrows to escape from drying winds. Birds, too, are attracted to the towns, because the thin groundcover and sparse grasses make it easy to see insects. Even ground beetles sometimes congregate in large numbers outside prairie dog burrows, and presumably go underground to winter below the frost line. According to the Bureau of Land Management, 10 species of amphibians, 15 species of reptiles, 101 species of birds, and 37 species of mammals use prairie dog towns as feeding grounds or shelter, and the population and diversity of insects is also greater there.

Like beavers, prairie dogs are rodents; they are short-tailed ground squirrels that measure from 13 to 17 inches and weigh from 1½ to over 3 pounds. They have short legs, long toenails, a keen sense of sight and hearing, and a complex social structure that includes a lot of nuzzling, grooming, and open-mouthed kisses lasting ten seconds or more. The kissing is often followed by snuggling and lying together. Individuals live from eight to ten years, usually in polygamous bliss in groups called clans or coteries, which are typically composed of one male, several females, and a half dozen young. During most of the year, the prairie dog's reproductive equipment is dormant, but starting in the fall the male's penis and the female's uterus begin to enlarge. They breed for two or three weeks in January or February, after which the reproductive organs of both sexes subside and they go back to nuzzling.

Prairie dogs are winsome creatures, but their survival strategy has a darker side as well. After the babies are born, many mothers go on an orgy of infanticide, killing and often eating the pups of others. Whether this behavior developed as a means of reducing competition for food or providing an extra shot of protein at nursing time, there are four to six weeks in the spring when hungry mothers turn prairie dog towns into dangerous places for newborns. When the pups that survive are able to care for themselves, the older prairie dogs move to the periphery of the town and expand its limits, leaving the youngsters in the safer, central burrows.

Prairie dogs are prey and act like it. Much of their day is spent watching for eagles and hawks, coyotes, bobcats, badgers, black-footed ferrets, and snakes, all of which rely on prairie dogs as food. When a prairie dog sees a predator, it utters a warning bark that sends all prairie dogs within earshot scurrying for their mounds, where they sit up and start barking too. In all, ten different calls have been identified: besides the warning bark, there is a specific warning for hawks, an all-purpose defense bark, a muffled bark, territorial call, disputing *chrrr*, chuckle, fear scream, fighting snarl, and tooth chatter. (The latter is occasionally used during disputes).

Prairie dogs are primarily vegetarians, and they are especially partial to wheat grasses and plants of the goosefoot family—though when food is low they will eat almost anything, from prickly pear cactus to burrs. They also enjoy grasshoppers and other succulent insects. Their caecum (which we've retained as our appendix) is as large or larger than their stomach, enabling two hundred and fifty tubby little prairie dogs to consume about as much food as a cow weighing half a ton.

When water falls on a prairie dog town, it falls on soil filled with what hydrologists call macropores—tunnels larger than a millimeter in diameter. Elsewhere, the entire soil profile, from the surface of the earth down to the water table, needs to be saturated before water can percolate down to the table freely, as it does from a buffalo wallow. But soil containing macropores doesn't need to be saturated, for macropores promote rapid transport of water through the soil. This process, called short-circuit bypass flow, violates a basic tenet of soil-water theory: prairie dog burrows allow water that would ordinarily not make it past the root zone to bypass the whole struggle and move directly to the water table. The soils in prairie dog towns are moister than soils in the surrounding area, and according to the hydrologists a higher soil moisture increases the total volume of water that percolates downward. Moreover, the high-intensity, short-duration rainfall that the Plains are likely to receive is the type of precipitation most apt to enter macropores and be rapidly funneled below the root zone. Clearly, the prairie dog population increased the amount of rainfall percolating down to the groundwater—and thence feeding the region's streams and rivers—as surely as the endless wallowing herds of buffalo did.

Here you have the great American grasslands, an incredibly complex system where Indians managed the fire, tens of millions of buffalo took care of the grazing, and billions of prairie dogs dug holes that, along with the wallows, increased the flow of water to the water table. The buffalo and the prairie dog, along with the beaver, made patches of habitat that ultimately raised the flow in the rivers, while the Indians' regular fires maintained the productivity of the grasslands.

Enter, from the right, Americans of European descent. The woodsmen who ventured West had never seen herds as large as those found on the grasslands, and the sight of so much meat on the hoof seems to have triggered a sort of blood lust. The buffalo were shot by the settlers for their tongues; the carcass and skin were thought to be worthless and were left to rot where the animal fell. The frontier diarist Nathaniel Henderson wrote of the Kentucky Barrens on May 9, 1775, "We found it very difficult to stop the great waste in killing. . . . For want of obligatory law, our game as soon as we got here, if not before, was driven off very much." Which is to say, no one owned the animals, so they were killed.

The bison herds of the Plains, though, were obliterated by market hunters for money. The economics were simple: salted buffalo tongues were bought for 25¢ each and sold in Eastern markets for 50¢. An undressed calf hide generally brought 50¢ (overcoats made of young bison fur were common and inexpensive), while those of adult animals in good condition cost $1.25. The bones, which were ground and sold to farmers as fertilizer, sold for $7 to $10 a ton. To maximize profits, many hunters dealt only in tongues, for it took far less effort to cut out a tongue than to strip off a hide. Men would often bring in two barrels of salted buffalo tongues without a pound of meat or a single robe. In a time when money was scarce and buffalo were plentiful, there were a great many hunters hauling barrels of salted tongues and packs of hides to steamboats on the Missouri.

John Charles Frémont, who surveyed the northern Plains for the U.S. Topographical Corps in 1839, noted that before 1836 a traveler crossing the Plains "would always be among large bands of buffalo." By 1840, the herds were shrinking noticeably. Frank Gilbert Roe, in his definitive study *The North American Buffalo*, quotes one Dr. Josiah Gregg, who observed in that year: "The vast extent of the prairies upon which the buffalo now pasture is no argument against their total extinction, when we take into consideration the extent of the country from which they have already disappeared; for, it is well known, that, within the recollection of our oldest pioneers, they were nearly as abundant east of the Mississippi as they are now upon the western prairie; and from history we learn that they once ranged to the Atlantic coast. Even within thirty years, they were abundant over much of the present States of Missouri and Arkansas; yet they are now rarely seen within two hundred miles of the frontier." In 1842, Frémont found the Sioux of the upper Plains *démontés*—undone—by the devastation of the buffalo. The following year, large villages of Sioux from the upper Missouri moved 500 miles southwest to the Platte in search of the dwindling herds. John James Audubon took a trip that year on the northern Plains and noted that the prairies were "literally *covered* with the skulls of the victims." And in 1844 Frémont noted that the buffalo occupied "but a very limited space . . . along the eastern base of the Rocky Mountains."

By the end of that decade, steamboats were bringing people up the Missouri as far west as Montana; the Santa Fe Trail led to New Mexico; and the Oregon Trail led from Independence,

Missouri, to the Pacific. In 1849, four thousand wagons and fifty thousand animals trekked west on the Oregon Trail in pursuit of California gold. This concentrated traffic was beyond the grasslands' carrying capacity, and the emigrants created a temporary desert, devoid of grass or game, in a long strip between Missouri and the mountains. Pioneers crossing the plains had killed buffalo whenever they could, and the constant harassment drove the animals away from the trails; after 1849, travelers crossing the Plains on the major trails rarely saw buffalo at all.

At the end of the Civil War, Texans drifted home to find millions of feral longhorns on the Texas plains, with ownership determined by whose land they grazed on. The first big cattle drive was in 1866, when over a quarter of a million steers were herded up to Kansas. The Kansas Pacific Railroad reached Abilene in 1867, and in four years more than a million Texas steers had been shipped to packinghouses in Kansas and Chicago, or to Iowa, Nebraska, Illinois, and Missouri farms for fattening on corn grown in fields carved out of the prairie. New cattle trails developed and their termini moved west to Newton, Wichita, and Dodge City, Kansas, to meet the Santa Fe Railroad. When stockmen learned that cattle could survive the cold winters, they moved them into Colorado, Wyoming, Utah, and Oregon.

As the railroads extended west and the longhorns pushed north, market hunters cleared the range of buffalo. Where buffalo were at all plentiful, every hunter killed between one thousand and two thousand during the hunting season, when the pelts were prime. The slaughter was greatest along the lines of the three great railways—the Kansas Pacific, the Atchison, Topeka & Santa Fe, and the Union Pacific. From 1872 to 1874, the railroads carried 1,378,359 hides, 6,751,200 pounds of meat, and 32,380,050 pounds of bones to market. On the Santa Fe route, it was said that one could walk 100 miles on the bloated bodies of slaughtered buffalo, and the southern buffalo range became a vast abattoir. Putrefying carcasses, many of them with the hide still on, lay thickly scattered over thousands of square miles of the level prairie. The remaining herds had become scattered bands, harried by hunters who now swarmed almost as thickly as the buffalo.

White hunters were not allowed to hunt in Indian Territory (the boundaries of which had been set as those of the present-day Oklahoma State in 1824), but they picketed the southern boundary of Kansas, and every herd that crossed north into Kansas was annihilated. Every watering hole was guarded by a camp of hunters, and whenever a thirsty herd approached, it was met by bullets. By 1874, the great southern herd was gone.

Before the construction of the Northern Pacific Railway, in 1880–1882, the only way to market the tongues and hides of the northern herd was to ship them down the Yellowstone and the Missouri Rivers to steamboats that brought them to rail-heads. Beginning in 1830, as many as a hundred thousand robes a year were sold from the northern herd, but the herd's denouement dates from the 1880s and the completion of the Northern Pacific.

As with the southern herd, it was the fact that a single hunter could destroy many hundreds of buffalo in a single day that erased the herd before the people of the United States were fully aware of what was happening. The Plains were so immense that few were able to imagine the great herds as finite. The Indians had believed that buffalo streamed perpetually from a cave deep in the prairies, and the buffalo hunters themselves had no idea that the great beasts were gone: they geared up in 1885 and went out right on schedule to find . . . no buffalo. In 1887, the zoologist William T. Hornaday wrote, "Twenty years hence, when not even a bone or a buffalo chip remains above ground . . . , it may be difficult for people to believe that these animals ever existed in such large numbers." And so the wallows slowly began to disappear.

While the herds were being slaughtered, the Plains Indians were being destroyed by gunfire, disease, and alcohol, as well as by the loss of their herds and their land. The U.S. military had turned its attention to emptying the Plains of Indians after the Civil War. The tribes' traditional religious, social, and governmental patterns were shattered, while the telegraphs and railroad allowed the federal government to move men and matériel wherever it was needed to quash the latest rebellion. Tribe by tribe and band by band, the Indians were defeated and moved onto reservations, where they were given hoes and seeds and told to become agriculturists. And so the annual burnings of the grassland ended.

The most obvious use of grasslands cleared of buffalo and Indians was to raise cattle, which could be taken by rail to be sold in the East. The only need was capital, which was supplied by Eastern dudes and British investors. The West had the grass and the cowboys; the East had the capital; the railways linked the two together. So grazing, at least, continued on the grasslands—but cattle and buffalo have important behavioral differences. Cattle have to be fed in winter, when they look around stupidly for hay on the surface of the snow. The buffalo found their own forage, scraping away the snow with their heavy heads and hooves. The grazing patterns of the two species are similar—if anything, buffalo are even more single-minded in their selection of grasses over forbs—but cattle and buffalo behave differently when they drink. Buffalo, being wild, don't linger streamside: they come down to the water, drink, and then leave. Cattle, long domesticated, just lounge around by the water, churning the slender green riparian edge into a muddy wasteland.

Like all edges, the riparian edge is the most productive fraction of its ecosystem, and streams that flowed clear when the buffalo drank from them grew muddy and rank under the clumsy hooves of cattle. Fish that once hid in the cool shadows along the banks lost their shelter, stream temperatures rose, and fish populations dropped along with the oxygen content of the water. Silt covered the gravel beds where fish had spawned; there were fewer places for frogs to hide from predators; the stream biota was simplified; and more soil washed into the water.

Prairie dogs, which are hard to shoot but exceedingly easy to poison, were the last to come under attack. Horses were said to break their legs galloping across prairie dog towns, and so were cattle, but the prairie dog was classified as a pest on account of its appetite, not its tunnels. Because they keep the vegetation in their towns closely clipped, it was assumed that they competed directly with sheep and cattle for forage. In 1901, the Yearbook of the U.S. Department of Agriculture described them as an evil that reduced the number of cattle an area could support by 50 to 75 percent, and the USDA provided detailed instructions on how to remove them with strychnine, at a total cost of 17¢ an acre or less. In 1920, the USDA Yearbook stated that prairie dogs cost ranchers $300 million annually by "selecting the most productive valleys and bench lands for their devastating activities"—never wondering why these lands were the "most productive." A concerted chemical assault was launched: that year, 132,000 men festooned 32 million acres with poisoned grain to improve the grazing for cattle.

Routine poisoning of prairie dog towns has continued unabated. Today zinc phosphide has supplanted strychnine as the poison of choice, and prairie dogs are still classified as pests. On public land, poisoning is attended to by the U.S. Forest Service, the Bureau of Land Management, and the National Park Service, while in many states prairie dogs on private land are "managed" by county weed and pest boards, which poison the prairie dog towns and attach the costs to the landowners' tax bills.

Ironically, recent research has shown that prairie dogs and cattle have a mutually beneficial relationship. Prairie dogs on the mixed-grass prairie need the help of large grazers to keep the plant cover low, for safety. When cattle are fenced out of a prairie dog town, the colony is unable to maintain a short-grass plant community, and the taller grasses hide predators that soon overwhelm the colony. For their part, cattle grazing in prairie dog habitat gain more weight than cattle grazing elsewhere on the Plains, thanks to the higher protein content of the tender clipped grasses. Forage quantity may be reduced, but its quality is improved, leaving at worst a neutral outcome. And it is also thought that the higher water content and organic content of the soils in prairie dog towns make the grass grow better there.

Nonetheless, the distaste of ranchers for these humble soil-making, groundwater-replenishing rodents lingers. They are tolerated only on a small fraction of their range on public land, and rarely go unpoisoned on private land. Prairie dog towns may have once covered a few hundred million acres; they now burrow in a total of about two million acres.

Without regular fires, the productivity of the grasslands declined. Without prairie dog towns to hide in and feed in, the snakes, box turtles, toads, and tiger salamanders, the cottontail rabbits, skunks, prairie chickens, and dozens of other populations began to decline; and the eagles and hawks, coyotes, foxes, badgers, and black-footed ferrets had less prey to feed on. As the biological diversity declined, the old buffalo wallows and prairie dog tunnels started filling in. The rate of groundwater recharge dropped, there was less underground water to feed the streams and sloughs, and slowly, slowly, across the Plains, the waters began to recede.

SIX

PLOWING THE PLAINS

As the natural system was being dismantled piece by piece, the prairies began to be plowed into fields. Plowing the prairie was impractical with Eastern plows: the sod stuck to the moldboard, and the farmer had to stop every two or three steps to knock the dirt off. John Deere's steel moldboard plow, first manufactured in 1843, was said to cut through sod like a hot knife through butter. The deep, rich, stoneless soil of the tallgrass prairies, with low but adequate rainfall, was among the most productive agricultural land in the world. It is estimated that less than 1 percent of the tallgrass prairie survives in its original state, mostly in fragments of graveyards and forgotten roadsides. Over time, it has been nearly completely plowed and planted with crops of domesticated tall grasses: spring wheat and corn.

The land was so level that heavy ox-drawn wagons with iron-bound wheels could travel across it without benefit of trails. With no forests, there were no tree roots to grub before the land could be plowed, and furrows could be laid a mile long. For the early pioneers on the Plains, the absence of trees also meant that the only building material was earth, and the only fuel was slough hay and buffalo dung. To secure shelter, new settlers would dig a hole in a suitable bank, lay ceiling poles across the top, and roof this den with reeds, sticks, and then sod. Leaky, dark, smoky, and warm, dugouts were small and so hard to keep clean that they literally drove women crazy. But they were temporary shelters, generally occupied only for the first year or two, while a sod house was being built.

Sod bricks were made in the fall, when the grasses of the prairie were woody. A plow would cut about an acre of sod into strips roughly 3 inches thick. The sod strips were cut into bricks that were laid sod side down: a house wall was generally three bricks thick, with the bricks staggered so that the joints didn't line up. Sod walls settled during the first year or two, so the window and door frames were built with a 6-to-8-inch gap above each frame, which was filled with rags and paper. As the walls settled, the gaps gradually closed, until the house was snug. Warm in the winter and cool in the summer, sod houses could be relatively elegant, too, with a kitchen, parlor, and several bedrooms. Earthen dwellings disappeared quickly, though, when the railroad crossed the country and a ready-made wood frame house could be ordered by rail from either coast in an easily assembled kit.

In the 1860s, the limitless prairies, largely owned by the United States government, began to pass into private owner-ship. The Homestead Act of 1862 allowed all citizens over twenty-one to have 160 acres of land, provided they lived on, worked on, and improved the land over a five-year period. Between 1862 and 1872, railroad companies received rights to twenty sections of unoccupied public land for each mile of track, totaling 180 million acres—about 10 percent of the con-tiguous United States—most of which was in the Great Plains. Finally, the Timber Culture Act of 1873 allowed settlers to claim 160 acres if they had planted 40 acres of it in trees. The trees fared so poorly in the hot summer winds west of the 97th meridian that the number of treed acres required was soon revised downward to 10.

The land was relatively easy to acquire, but the farming was hard. Farmers on the tallgrass prairies contended with low but generally adequate rainfall, which came in torrents rather than showers, and also with high winds, high summer temperatures, and low humidity. In late winter and early spring, winds tore at the fields; in summer, the hot, dry winds sucked the moisture out of the plants and the soil. But the earth was rich, the rain-fall was usually sufficient to grow lush crops of corn and wheat, and so the perennial grasses of the tallgrass prairies were replaced by annual grains.

On the mixed-grass prairie, farming was more uncertain. In addition to the winds and the hot, dry summers, the rainfall was too low to reliably produce a crop of spring wheat or corn, and the streams were too scarce and intermittent to provide irriga-tion. Land bordering waterways was relatively arable, but even the proximity of a stream was no insurance against low rainfall: the farmers upstream would dam the creeks to generate a con-stant summer flow to their fields and stock, while farmers downstream were left with a bed of dry rocks. The doctrine of riparian rights had been imported from the East, where landowners who lived next to a stream had the right to use the water. Riparian rights allowed a tiny handful of landowners to monopolize the few manageable rivers in the West, while the rest of the settlers learned painfully that without irrigation, 40 acres in Illinois or Iowa produced more grain than 160 acres in western Kansas.

With insufficient water above ground, farmers brought water up from below with windmill-driven pumps, generally with cylinders 10 inches in diameter. These were enough to fill a reservoir that could water only 30 head of cattle or 5 acres of cropland, on which a family could barely survive. The drier, harsher climate of the mixed-grass prairie required new farm implements, new crops, and new skills.

Dryland farming techniques were introduced to the settlers of the mixed-grass prairie by a group of Mennonite farmers who had emigrated from Russia beginning in the early 1870s. They were Germans originally: Catherine the Great had brought them to Russia a century earlier to farm the Caucasus. Their rights were abolished in 1871, persecution began, and literally tens of thousands of Mennonites had moved to the Plains by the 1890s. With them they brought hard red winter wheat, which required less moisture than spring wheat. They had had long experience with dryland farming, and allowed their acres to lie fallow in alternate years, in order to build up subsoil moisture from two years' worth of rainfall. The Mennonites were joined by Scandinavians, Germans, Ukrainians, and many others who wanted to plow their own land. Hard winter wheat—mixed with the seeds of the Russian thistle, or tumbleweed—came to dominate all crop production in the mixed-grass prairie, and the tumbleweed bounded across the Plains.

Then, from about 1878 to 1887, a miracle happened. The rains came, year after year, and farmers moved farther and farther west, concluding, with U.S. land office agents and the public at large, that turning the sod had changed the climate from dry to moist. The rain was said to follow the plow, and it was also widely held that the iron and steel rails of the railroads and the metal in the telegraph wires had altered the natural electrical cycles in a dry region, inducing rain.

During the rainy decade of the 1880s, fields of short winter wheat and drought-resistant dryland grain sorghums spread over the mixed-grass prairie, and farmers were using increasingly large machinery. The hours of labor required to plant, tend, and harvest crops dropped with the purchase of each new piece of equipment, while record production led farmers to believe that all the land had needed in order to flourish was the white man's plow. Optimism notwithstanding, the increase in rainfall was temporary, and bad weather returned with a vengeance. The blizzards of early 1886 were so fierce that settlers lost their way between house and barn and wandered the Plains until they froze. The entire Eastern- and British-financed open-range cattle industry died that spring, along with most of the cattle.

A drought began in 1887 and continued through 1890; two rainy years were then followed by five more years of drought and hordes of grasshoppers that "ate everything but the mortgage." There was as yet no dust bowl, because enough perennial groundcover remained to hold the soil in place, but entire communities were hammered into the ground, and disillusioned settlers told tales about a Kansan mule standing in a field of corn: the weather was so hot that all the corn started to pop, and the mule thought it was a blizzard and froze to death. Some farmers left the Plains with less than they'd come with, moving back East or on to the West Coast and festooning their wagons with such slogans as "In God we trusted; on the Plains we busted."

The harsh climate and difficult conditions did not deter many of the land-hungry immigrants, though, and farmers kept coming to plow the mixed-grass prairies. By 1900, half a million families had settled in the region; by 1910, farmers had claimed much of the southern Plains. Meanwhile, farming was so uncertain on 160 acres of mixed-grass prairie that the five years of homestead residency came to be called "the period of starvation." A new homestead act passed in 1912 addressed the realities of Plains agriculture by changing the required residence period to three years.

By this time, integrated mechanization—gasoline-powered tractors and trucks, and combines that reaped and threshed in one operation—allowed a single dryland farmer to manage thousands of acres. But all that machinery placed new cash pressures on the farmer, who had to pay off the mortgages on the equipment. Unfortunately, the early combines could not harvest most sorghums (which require less moisture than wheat), so farmers who mechanized were locked into wheat production and generally did not diversify their crops until the equipment was paid off. The emphasis on wheat meant financial ruin when the rains were sparse; but for the most part there was enough moisture, the wheat grew, and farmers expanded their acreage.

The years between 1914 and 1929 have been called the Great Plow-Up. Those were the halcyon days of virgin fields, of a burgeoning European market for wheat after the chaos of the First World War, and of fine new machinery. The most popular plow was the one-way disk plow, which left a smooth, pulverized soil that needed no additional harrowing. The organic matter in soil breaks down more quickly if there are no clods, and the new plow helped to produce bumper crops of wheat, but it also contributed to the collapse of the soil structure. The prairie soils had originally been deposited by the wind, which is why they were stoneless. This windblown soil, called loess, is characterized by flocculation—that is, the soil particles cling together, like clumps of caviar, and these clumps create interstitial channels that allow water to filter down to the subsoil. Long drought destroys the soil's ability to flocculate, particularly if the soil is plowed and planted again and again by farmers who have a lot of expensive equipment, plenty of land, no rain, and are desperate to pay their bills. Where there is no flocculation, the soil breaks down into tiny dust particles that are easily swept into the air by the high winds on the prairie.

In the early 1930s, when much of the best land in the mixed-grass prairie had been plowed and planted, wheat prices dropped, falling from 99¢ a bushel in 1929 to 34¢ a bushel in 1931; and the rain did not fall. Drought, combined with the Depression, hit the Plains farmers hard. In order to increase their payoff whenever it finally rained again, they plowed more and more land. In the Kansas portion of the dust bowl alone, plowed acreage increased by 25 percent during the 1930s.

The dust rose because there was not enough rain for ground-cover to establish root systems, and plowed fields had nothing holding the soil down in the windy months of February, March, and April. Crops failed and land was abandoned. On plowed fields, it takes 30-mile-an-hour winds to start moving loess into the air, but once the soil starts to blow, a wind velocity of less than 15 mph is enough to sweep more into the air.

At first, most of the dust storms were local, and typically confined to sandy lands, where crop failure had left the ground bare. Dust storms became more common in 1933, when minimal rainfall and higher-than-average winds brought widespread wind erosion. The land to the depth of a plowshare was gone in some regions, and the windblown soil piled up in others. Headlights and house lights had to be turned on in the afternoon, the sunlight turned violet-greenish, and cattle huddled against the blowing dust as if it were a blizzard. In 1934, prairie dust dimmed a swath of the East Coast, from Washington D.C. to New York City, and dust fell even on ships 500 miles out at sea. For the first time in their lives, many Easterners tasted Plains soil.

In 1935, in the Texas panhandle, the dust storms took the form of black blizzards. Cattle suffocated, their lungs caked with dust. A Kansan housewife wrote, "All we could do was sit in our dusty chairs and gaze at each other through the fog that filled the room and watch the fog settle slowly and silently, covering everything . . . in a thick, brownish gray blanket." For some reason, the jackrabbits began multiplying unchecked in Colorado and Kansas, and hundreds of thousands of rabbits competed with cattle for the scarce grass. Farmers armed with sticks, bats, and old tool handles organized rabbit drives on Sundays after church, whenever the dust didn't blow.

Before the 1930s, many farmers had believed that a little wind erosion was a good thing. Wind was thought to mix the soil and help maintain fertility. During the long drought, they reassessed their stance on wind erosion and changed their agricultural practices accordingly. Smooth and clodless fields had allowed the dust to blow unchecked, but when fields were "listed"—or deeply furrowed—clods remained, and the tiny plants could gain strength and size behind each furrow to withstand the onslaught of the wind, increasing the likelihood of establishing a groundcover. Moreover, chiseled fields opened the soil so that rain could penetrate below ground and keep the subsurface moist.

Starting in the spring of 1935, the Federal Emergency Relief Administration spent a total of $2 million to pay farmers for the time and fuel it took to list their fields; by the summer of 1937, nearly 10 million acres had been listed, and wind erosion was curbed. The Soil Conservation Service found that contouring and terracing fields also increased soil moisture and crop yields. Strip cropping, in which strips of drought-resistant sorghum alternated with strips of wheat, was promoted to reduce wind erosion. When the Plains farmers combined rough tillage practices with contour furrowing, contour terracing, and contour strip-cropping (with the contours at right angles to the prevailing winds), less soil blew away. In addition, nearly 20,000 square miles of "submarginal" land was bought or appropriated by the government and taken out of tillage. (This program, which ended in 1947, established the Kiowa Grasslands and the Cimarron National Grasslands in Kansas, among others.)

In the spring of 1938, the rains returned and the dust storms decreased in frequency and intensity; by 1939, more rain had ended the threat of desertification, and agriculture began to boom once again. Meanwhile, the dust bowl had led to a great discovery: water was there on mixed-grass and short-grass prairie, if you dug deep enough. The Ogallala aquifer, lying roughly between the 105th and 100th meridian, ran from the southern edge of South Dakota deep into the Texas Panhandle, supplying from beneath the ground what was missing up above.

An aquifer is a layer of buried rock in which the interconnected gaps fill with water. If the gaps are big and there are a lot of connections, the water will flow easily; if they are small and poorly connected, the flow will be restricted. The rate of flow of an aquifer is determined by its permeability; the amount of water it can hold is measured by its porosity. Basalt, an igneous rock, has plenty of big gaps but these tend to be discrete; basaltic formations are highly porous but nearly impermeable; a cracked block of granite has high permeability but low porosity. The best aquifers are both porous and permeable—usually buried beds of sand or gravel, or heavily fractured bodies of sandstone or limestone.

The Ogallala is a buried remnant of a vast alluvial plain that stretched east from the Rockies at the end of Miocene time, five million years ago, when the Rockies were the size of the Himalayas. The climate was cooler and wetter then, and great glaciers in the high peaks wore the mountains down, creating porous deposits of gravel, rock, and sand that washed down onto the plain. As the climate gradually grew warmer, the glaciers retreated, erosion slowed, and the Rio Grande and Pecos Rivers eventually cut through the alluvium faster than it was being deposited. These deepening rivers diverted the Rockies runoff, which once had flowed to the Ogallala deposits, cutting the Ogallala off from its recharge area. A drastically reduced level of rainfall led to a process called calichefication, in which soil particles become cemented together by minerals left behind as water evaporates. This process created a thick, watertight roof, like a potter's glaze, over the saturated gravel bed: the Ogallala water was trapped and preserved to the present day. Fifty to three hundred feet below the surface of 174,000 square miles of fertile but dry mixed-grass prairie lies the equivalent of Lake Ontario.

By the end of the dust bowl, irrigation technology had progressed considerably from the homemade windmill pumps, which could not raise water any higher than 30 feet. First the centrifugal pump and the gasoline engine made the windmills superfluous; then less expensive well drilling and casing made deeper wells economically feasible. Finally, inexpensive plastic, rubber, and concrete pipes became available after the Second World War, reducing the labor and costs of an irrigation system. Up came the water for stock troughs and the vegetable garden and fruit trees, and soon for cash crops of corn and milo and wheat and cotton. In 1935, West Texas farmers pumped 8.4 billion gallons from the Ogallala; fifteen years later, they were using more than 350 billion gallons annually and irrigating 3.5 million acres of farmland. Today, the Ogallala irrigates 16 million acres. The aquifer holds over 3 billion acre-feet, or a quadrillion gallons of water, and this water is being mined at a very high rate. An estimated half of the usable water has been spent; current consumption patterns should take care of the rest in perhaps thirty years. As a direct result of pumping the Ogallala, much of the mixed- and short-grass prairie is now reliably plowed for crops.

The perennial grasses that once covered the prairies held the soil in place when the wind blew, and caught the rain as it fell, allowing it to drip from leaf to soil and—if it managed to escape the root systems—down to the water table. Irrigated fields of annuals, however, leave the ground bare for part of the year, and without the blotterlike plant cover, runoff increases. When rain falls on plowed land, topsoil is swept along with the flow. Muddy torrents gully the ground, and the Mississippi becomes laden with silt as the topsoil is swept to the sea.

The short-grass prairie, which does not get enough rainfall to support dryland farming and does not lie over a major aquifer, could not support small farmers. As a result, its agricultural development followed a significantly different path from that of the tallgrass and mixed-grass prairie. Although Indians and Hispanics had irrigated in the short-grass prairie for centuries, the Mormons laid the pattern for nineteenth-century irrigation in the West.

The Mormons settled in Utah's Great Basin in 1847 to practice their religion in peace. Their leader, Brigham Young, declared all water, timber, and mineral rights as common property, in contrast to the rest of the West, where landowners abutting the waterways controlled the water. The earliest Mormon settlements were in valleys at the base of the Wasatch Mountains, where rainfall rarely exceeded 15 inches a year. (The mountains had 30 to 40 inches of precipitation, which mostly fell as snow.) Crops had to be irrigated, and the Mormons managed this in characteristic fashion. Pooling their labor, equipment, and work animals under the direction of their local bishops and according to the precepts of the church elders, they built earthen dams and dug irrigation ditches that soon watered tens of thousands of acres of newly turned short-grass prairie. Tithing provided the income base, and the church provided the authority to mobilize large amounts of labor; this happy marriage of labor and capital controlled by a central authority allowed water to be moved on a grand scale. Thousands of miles of canals were dug, hundreds of dams erected, and the Great Basin became the Zion the Mormons had long sought.

In the first decade of emigration, the Mormons built a hundred towns and villages based on irrigation, almost all in Utah. In the second decade, 135 more communities of irrigators were founded, not only in Utah but in southern Idaho, southern Nevada, and northern Arizona. In the third decade, 127 additional villages were established in Utah and Arizona.

Elsewhere, riparian rights prevailed, but this was not practicable in the West, for it allowed a few landowners to control a critical resource. When miners came West to dig the silver and gold from the mountains, they all needed streamside access to process their ore. The doctrine of appropriation—known colloquially as "first in time, first in right"—effectively made water a form of personal property, and the first person that diverted water from a stream—regardless of how far the water was diverted from its natural course—felt free to exploit it as long as the water still flowed.

Once "owned," water rights could be sold. The Desert Land Act of 1877 allowed settlers to claim 640 acres (a square mile) of the short-grass prairie, provided that the claimant built an irrigation system in three years. Although the land was ideal for light grazing, there was a moral imperative at work: the federal government felt that the highest and best use for fertile, temperate, level American land was crops. Hundreds of Eastern-backed water companies were formed to profit from the potential windfall, most of them in regions that (aside from the lack of rainfall) were well suited to agriculture: California's Central Valley, Nevada, Arizona, southeastern Colorado, and New Mexico. An irrigation company would build a dam or reservoir, dig canals and ditches, and sell water until the dam washed out or the canal construction became too daunting.

Almost all the private irrigation companies went bankrupt in less than ten years. At the eighth National Irrigation Congress, in 1898, a Colorado legislator likened the short-grass prairie to a graveyard littered with the "crushed and mangled skeletons of defunct [irrigation] corporations . . . which suddenly disappeared at the end of brief careers, leaving only a few defaulted obligations to indicate the route by which they departed." Private enterprise had failed to turn the arid West into a garden. By 1889, there were only 5,700 square miles of irrigated cropland west of the 100th meridian, and half of that was Mormon.

To encourage the states of the Far West to finance the construction of dams, reservoirs, and canals, the Carey Act of 1894 directed that the federal government deed up to 1,500 square miles of land to any state that would irrigate it. The result was disappointing: in fifteen years, only 450 square miles were irrigated and thus transferred to state control. And so the arid grasslands of the Great Plains and California reached the twentieth century relatively intact. The rivers were so large and the landscape so vast that neither private initiative nor state financing could build dams massive enough to tame the Western waterways.

Nonetheless, Americans were continuing to move West. Even during the "white winter" of 1886 and the years of drought that followed, the westbound trains were full. To support the growing Western population and prevent recurring disasters, the region needed irrigated agriculture. And so the Reclamation Act of 1902 was passed, throwing the full weight of the United States government behind the mammoth task of dam building.

~

ENGINEERING THE WATERWAYS

SEVEN

THE WATER OVER
THE DAM

The intent of the Reclamation Act of 1902 was to use public funds to irrigate the otherwise unfarmable deserts of the West. In 1903, six major irrigation projects were approved, and the Reclamation Service, which had been established by the act, began construction on the Truckee-Carson Dam in Nevada, to irrigate an area that was virtually uninhabited. In 1905, work began on the Roosevelt Dam, on the Salt River near Phoenix, Arizona. The dam was built with rock hewn from the canyon walls by a crew of Apaches not twenty years removed from Geronimo's band, and a few Mexicans and hoboes recruited from ranches and freight trains. When completed in 1911 at a total cost of $10.5 million, the Roosevelt Dam was the tallest rock masonry dam in the world. A solid wall bowed upstream against the river's current, rising 280 feet above bedrock. The Reclamation Service's ingenuity and resourcefulness and the graceful design of its dam astounded the world. The grasslands of the Salt River Valley, now assured an ample, inexpensive supply of water, were soon farmed, and Phoenix was on its way to becoming a desert metropolis.

After the Roosevelt Dam on the Salt River came the Arrowrock Dam on the Boise, which was even taller, the Elephant Butte Dam on the Rio Grande, and the Gunnison tunnel, dug through granite and shale to divert Colorado's Gunnison River into the Uncompahgre Valley. By 1919, twenty-six separate reclamation projects were in various stage of completion.

The newly irrigated lands were generally planted in alfalfa, wheat, cotton, or hay—all low-value crops that were cheaper to raise in nonirrigated regions. Except for California lettuce and oranges, crops grown in the desert did not make the farmers rich. The overhead was high, and it was more difficult to make a living growing irrigated crops in the short-grass prairie than nonirrigated crops in the tallgrass prairie. In the good year of 1917, for example, a Department of Agriculture study showed that the Corn Belt farmers in Illinois cleared an average of $870; in Chester County, Pennsylvania, the farmers made $789; but on the irrigated land of Utah's Salt Lake Valley, farmers made only $417.

Nonetheless, the concept of reclaiming a desert was a potent lure, and in the first decades of the twentieth century, some of the best engineering graduates in the country gravitated toward the Bureau of Reclamation (as it was renamed in 1923). Projects were financed by a fund initially made up of revenues generated by the sale of public land in the West, and construction costs were meant to be paid back gradually, through the sale of water to farmers.

As the size of the projects increased, so did the cost of the engineering and construction. But the farmers made too little money to pay much for water, and it was soon clear that the cost of building a dam was unlikely to be recouped through water fees. The difference between the farmers' payments and the actual cost of the Bureau's forty-year Columbia River project, for example, amounted to a 96.7 percent public subsidy by 1980. Dams were not built as moneymakers, though: they were built to bring farmers and civilization to unpopulated land. As a good landscape was one that lay plowed, planted, and stocked, so a good river was one that was dammed, channeled, and leveed. From an engineering standpoint, dams were wonderfully multipurpose construction projects, useful for hydroelectric power, flood control, and water supply as well as irrigation. Every canyon was a hole that was waiting to be filled, and water that ran to the sea without being used was considered wasted.

The dams generally had modest effects on the waterways until the 1930s. Floods were regulated and sediment was trapped, but the dammed rivers retained some of their natural character. Seasonal flow variation was less pronounced but nonetheless apparent; reservoirs were relatively small and affected the stream channel only for a limited distance. But in the 1930s the construction of large dams and the grouping of multipurpose projects within entire river basins became symbols of American engineering mastery over the landscape. The Tennessee Valley was the first fully integrated basin development in the country, and others soon followed. In 1936, the Bureau blocked the Colorado River with the Hoover Dam, which, like the Great Wall of China, is said to be visible from the moon with the naked eye. In Washington State, the Bureau concentrated on the Columbia River Basin. Eight dams had already been built on tributaries to the Columbia when, in the

1930s, the Bureau decided to dam the Columbia itself, on a glacial outwash known as the Grand Coulee. The Grand Coulee Dam was completed in 1942; ambitious projects were under way in the Central Valley of California; and before the century was half over, the American West had more and bigger dams on its waterways than any other region on the planet—and still the building continued.

Between government and private enterprise, roughly fifty thousand dams have been built in this country. Of these, a thousand or so are what engineers refer to as "major works"—truly gigantic constructions holding back rivers that were once thought to be untamable: the Colorado, the Columbia, the Snake, the Tennessee. There are sixty dams on the Missouri River and its major tributaries, and twenty-five dams on the Tennessee. Why on earth did we build so many?

With the adoption of river-basin planning in the 1930s, a single project typically included dams, canals, and irrigation works from headwaters to river mouth, across 1,000 miles of terrain. During the Roosevelt and Truman administrations, several omnibus river bills authorized dozens of dams and irrigation projects at a single stroke. Economics mattered little, for while the irrigation projects might never pay off, the hydroelectric dams within the same river basin could generate the revenues to cover the losses.

Over a period of eighty-six years—from 1905 to 1991—the Bureau of Reclamation and its predecessor built 339 reservoirs, 154 diversion dams, 7,670 miles of irrigation canals, 1,170 miles of pipelines, 270 miles of tunnels, 267 pumping plants, and 52 hydroelectric power plants. With the help of $18 billion in capital outlays, over 14,000 square miles of farmland received water, and the West bloomed and prospered. Water made much of this land intensely productive, and advocates of reclamation saw public irrigation as a miraculous font of free riches. There were, however, great costs.

~

Mammoth dams are able to exert total control over the down-stream river. The waterways are manipulated for power, flood control, and water supply, and the result is a truly engineered river system, whose out-of-season high flows don't correlate with seasonal rains. Natural extremes of flow, water temperature, and sediment transport are eliminated. Instead, unnaturally rapid flow fluctuations or sudden periodic flow changes are often superimposed on a constant background discharge: the low flows are higher, the high flows are lower, and the cleansing and nutrient-filled annual flood pulse is eliminated.

When the Hoover Dam went up, fish that had evolved to survive in the widely fluctuating and muddy flows of the Colorado lost their competitive advantage. The muscled humps of the humpback and bonytail chubs, designed to allow these fish to hold position or make progress against currents that swept most fish away, were suddenly superfluous. Eight dams on the lower Colorado and an aggressive dredging program to redirect the riverbed have contributed to the extinction or threatened status of eight species of fish native to the Colorado. As a kind of compensation, channel catfish and rainbow trout were introduced, and have prospered.

Water takes much longer to move through a dammed river system, and nearly all the sediment and organic matter carried along by the water is permanently stored in the reservoirs. There is less organic matter in the water to fuel the aquatic ecosystem; worse, since natural river channels are maintained by a dynamic equilibrium of sediment deposition and erosion, water without sediment can trigger major riverbed erosion. When dams retain silt, erosion is accelerated for dozens of miles downstream. Below the Hoover Dam on the lower Colorado, for example, millions of cubic yards of sediment were scoured from the channel for 100 miles, and the channel slope was noticeably reduced. It took about twenty years after construc-

tion for the channel bed to be fully armored and stabilized—
that is, transformed into a boilerplate of bedrock and boulders.

The water that flows in a stream is well mixed and well oxygenated. When water rests in a reservoir, it starts to stratify. Since
cold water sinks and warm water rises, reservoirs generally
stratify all summer long. Phytoplankton proliferate near the surface of the reservoir, releasing oxygen and keeping the water
saturated with it. Little mixing occurs, and the sunlight does
not penetrate below the top few yards. Dead phytoplankton
settle to the bottom and are consumed by bacteria, which use
oxygen to process the corpses, so water at the bottom of the
reservoir is low in oxygen, high in organic matter, and high in
salts as well. When autumn comes, the reservoir overturns, as
the water on top chills and sinks and the water on the bottom
becomes comparatively warmer and moves to the surface. The
high-nutrient, high-salinity, low-oxygen water is now on top,
while the low-nutrient, low-salinity, highly oxygenated water is
on the bottom of the reservoir.

A dam is designed to release water either from the top or the
bottom of its reservoir. Dams with deep-outlet discharges create streams with high nutrient levels, high salinity, and low oxygen during summer months, while in the fall these same
streams are fed by nutrient-poor, low-salinity water with plenty
of oxygen. During the summer, surface-outlet discharges create
streams low in nutrients and salinity, and in the fall they create
streams high in salinity and nutrients. Below a reservoir, thermal and chemical gradients in the river will persist for dozens
of miles downstream. Silt builds up in whatever gravel beds
remain, and the native fish are edged out by introduced species.
Water quality shifts in odd ways, and the daily temperature variation in the stream disappears. Water temperature in a stream
varies daily, by perhaps 7°F in the winter and 20°F in the summer. Below a reservoir, the daily temperature variation, if there
is one, is generally less than 1°F.

These changes seem subtle to those of us who live on land, but to aquatic species they are dead obvious. Water temperature, for example, triggers egg hatching in many aquatic insects. Several species of mayflies require three specific temperature regimes to hatch: near-freezing water triggers egg development; a rapid rise in the water temperature precedes hatching; and minimum temperatures of about 65°F over several months stimulate nymph maturation. The constant temperatures below a deep-release dam may not provide the thermal cues necessary for an insect species to complete its life cycle, leaving few larvae or insects for the fish to eat.

When a watershed is dammed and lakes and reservoir-fed rivers replace the swift, turbulent, highly oxygenated water of a free river, the riverine species will begin to fade away. In time, the entire aquatic community of a dammed river will begin to resemble that of a lake ecosystem. Moreover, when rivers are dammed, the fish that mature in the ocean and swim inland to spawn in the rivers begin to disappear.

The Pacific salmon are among the last of the great anadromous fish runs, but there once were many fish species on both coasts which journeyed inland to spawn. In his 1705 *History of Virginia*, Robert Beverly states that "herrings and shad come in such abundance in the brooks of Virginia that it is almost impossible to ride without touching them." In 1851, William Stack reported that his father had seen "shad so thick in the Merrimack River that you could not put your hand in without touching some of them, and yet there were more alewives than shad." And when the eels ran in the streams of Onondaga, wrote Jesuit diarist David Zeisberger in the 1780s, a man could spear a thousand in one night. Eels are catadromous, spawning in the sea and maturing in rivers that empty into the Atlantic Ocean. When it's time to breed, eels lay on migratory reserves of fat totaling up to a third of their body weight, and swim thousands of miles to the deep of the Sargasso Sea, which lies south of Bermuda. There they lay their eggs, and die soon after. The young females return to fresh water; the males stay in salt or brackish water all their life.

Striped bass 5 feet long and weighing as much as 100 pounds once entered the streams of the mid-Atlantic states in great abundance, along with the shad, the alewives, and the Atlantic salmon; and the sturgeon were once so regular in their inland expeditions that the full moon of August was called the sturgeon moon. The largest of all freshwater fish, some sturgeon live for two hundred years, measure 20 feet long, and weigh close to a ton. (The sturgeon of Russia's Lake Baikal are said to reach 2,200 pounds and an age of three hundred years—estimated from annual growth rings on their eardrums—although it seems highly unlikely to me that a fish could survive from the reign of Peter the Great to the present.) Some species of anadromous fish die after a single spawning, but the sturgeon returns to the sea and makes forays inland to spawn every few years. Sturgeon eggs are nice as caviar, and their swim bladders were once used to make isinglass windows (that roll right down, in case there's a change in the weather).

The sturgeon of the Great Lakes were so large that the spawning runs were easily obliterated by commercial fishermen at the turn of the century. As early as the late 1700s, overfishing began wiping out the Atlantic salmon as well. The survivors are much smaller today than they were two hundred years ago. An early writer reported that 40-pound Atlantic salmon were occasionally taken on the Grand Codroy River in Canada. In this generation, the average salmon caught there tips the scale at 4 pounds, with the largest fish, year after year, weighing in at less than 25 pounds. But the anadromous species that commercial fishing didn't take care of, dams did. The rivers of the East powered the early Industrial Revolution, with dams built by entrepreneurs running mills of all sorts. Dams proliferated on the smaller streams, and in the first half of the 1800s industrialists commandeered the larger streams, while lumbermen dammed the rivers to create high flows for enormous log drives. The New England states passed laws to protect the salmon to no avail, since the fish could not swim past the dams. By the 1870s, the Atlantic salmon had virtually disappeared, while there were 433 laws on the Maine books to control and preserve the fisheries.

In 1890, perhaps haunted by the recent disappearance of the Atlantic salmon, Washington State's first legislature passed a law requiring that fish-passage devices, such as fish ladders, be built on dams "wherever food fish are wont to ascend." Federal fisheries laws were also passed in the last two decades of the century, and these also required that dams permit some way for migrating fish to pass. Nonetheless, by 1940 eight unladdered dams had been built on the Yakima River in Washington's Columbia Basin, and they had decreased the annual salmon run from six million to nine thousand fish. The Grand Coulee Dam effectively closed more than 1,000 miles of spawning streams in the upper Columbia Basin, destroying the legendary "June hog" King salmon run in that region.

Salmon, the most studied of the anadromous fish, are beautiful swimmers, who glide effortlessly on unseen currents and can accelerate faster than a car for short distances. The Chinook, or King salmon, is the largest of the Pacific salmon, and will occasionally weigh in at 125 pounds, with a 5-foot-long body; the other four species range down to 20 pounds packed into a 2-foot torpedo. And they used to be numerous. The Columbia River once had annual runs of an estimated sixteen million salmon.

Young salmon live in the ocean, feeding on zooplankton, small fishes, and squid. They follow the prevailing North Pacific currents in sweeps of 1,000 miles, traveling regularly each year in migratory patterns that depend upon the currents, the climate, the contours of the ocean bed, and possibly even the sun and stars. Salmon have good eyesight and can also detect pressure waves, using currents to navigate; they may even navigate using the earth's magnetic field. Fish migration is now believed to be based on exploration and learning, and as new habitats arise, the fish rapidly exploit them.

When a salmon is about four or five years old and has traveled 10,000 miles or more in its circuits of the great oceanic grazing grounds, it puts on weight and heads inland to spawn. Some salmon spawn in the spring, others in the fall; some swim as far as 2,000 miles inland to reach their natal stream. Salmon, like many fish, follow their noses. They distinguish between species and sex of other salmon by scent, and use olfactory landmarks to determine their location. They remember the smell of the vegetation and aquatic residents of the stream they grew up in, and smell their way back to breeding territory.

Although the salmon's life path was never easy, it has become unspeakably difficult to be anadromous today. If a salmon needs to ascend (for example) the Columbia to find its natal stream, it has to start with the Bonneville Dam. Fish mill about at the bottom of the dam waiting their turn at the fish ladders, bottlenecked in the odd-quality water straight from the reservoir. At the top of the Bonneville ladder, they are confronted with 40 miles of warm, placid water, very different from the cool, oxygen-filled waters of the rapids and river currents it replaced. After the Bonneville Dam comes the Dalles and then the John Day and the McNary—each a major barrier to fish migration and each a major lake to swim through. Next come the Priest Rapids Dam, the Beverly Dam, and the Rock Island Dam, for the Columbia is no longer a river but a series of slackwater lakes.

Every dam that went up inundated a stretch of spawning stream on one side and eroded the gravel on the other. When too many dams are built on a river and too many downstream miles of riverbed are scoured, there are too few miles of gravel beds left for riverine fish to make nests in. The salmon, having made the journey, will dig their redds one on top of the other if necessary, each fish laying and fertilizing perhaps five thousand eggs.

Salmon hatch after two months of incubation in their gravel nests. The hatchlings live in the gravel until the yolk sacs attached to their bellies are consumed, after which they are recognizable as little fish. The fingerlings hide behind logs and under ledges, grab their food quickly, and retreat for cover. They swim near the bottom of the stream, where the current is weakest, traveling alone and conserving energy by resting in eddies. They come to know the stream's hiding places and its predators, and when the water runs low the young salmon will dive into the gravel rather than swimming unprotected into the current. They are surprisingly skillful at making their way through the gravel substrata, and have been known to follow underground watercourses into wells and springs, making it seem as though salmon fell from the sky. The fingerlings live in their river for as long as a year before becoming smolts, a silvery transformation that allows them to survive in salt water and immediately precedes their swim to the sea. Which is when things get extremely dicey.

A highly stratified reservoir can have surface temperatures that are lethal to juvenile salmon, while the cooler subsurface water is usually low in oxygen. Reservoirs are fine homes for lake fish, but they can become impassable barriers for juvenile salmon. Once the smolts get through the reservoir, they must pass through the hydroelectric turbines, which can be a bruising experience. In low-flow years, they may have to go through the turbines of seven or eight dams before they reach the ocean, and as many as 95 percent of the smolts will be macerated en route.

In high-flow years, the juveniles may be able to avoid the turbines, but the water that goes over the dam's spillway captures air and plunges it to substantial depths, mixing too much nitrogen into the water. A young salmon swimming in water supersaturated with nitrogen can die of the fish equivalent of a massive attack of the bends. Nitrogen bubbles form under their skin, their eyes bleed, sometimes their internal organs explode. In 1970, when the nitrogen levels in the water of the Columbia and Snake Rivers reached 143 percent of air saturation, the National Marine Fisheries Service estimated that 70 percent of downstream migrant salmon and steelhead trout were killed by nitrogen supersaturation before they reached the ocean.

The Pacific salmon's habitat has also been degraded by deforestation. Streams in deforested land are warmer, and the quiet pattern of pools and riffles created by fallen logs disappears, reducing food sources for the fingerlings. The salmon coevolved with the beaver, and until John Jacob Astor made his fortune in beaver skins from the Columbia River Basin, beaver ponds were part of the salmon's early life and part of its success as well. It is not surprising—fish ladders and legislation aside—that the salmon population started to fall off rapidly as the Pacific Northwest was dammed.

The solution to this complex environmental problem was technical: build hatcheries. The first salmon hatchery in the United States was built in 1871 on the Columbia, and between 1900 and 1930 the number of fry annually released by the state increased from twenty-five million to ninety million. So great was the government's faith in hatcheries that when a dam was built on the Olympic Peninsula's Elwha River in 1915, the Elwha hatchery was opened in lieu of emplacement of fish ladders, and in the next few years seven more hatcheries were built to compensate for the fish runs lost to dam construction on the White Salmon, Chehalis, and Elwha Rivers. Initially, two million eggs were collected annually to hatch in tanks, but within a few years the pool below the Elwha dam was empty of fish. The Washington Department of Fisheries abandoned the Elwha hatchery in 1922.

In the 1930s, Canadian fisheries biologists showed that substantial releases of hatchery sockeye increased neither the commercial catch nor the number of fish spawning in the wild. Canada closed all of its Pacific salmon hatcheries, and a number of American hatcheries on the West Coast were closed. But the Washington Department of Fisheries expanded its network. In 1958, the first of twenty-five Washington fish farms was built. By 1966, the program, which used natural lakes to raise large numbers of confined fish, was abandoned, because almost no farm fish were coming back to spawn. But to prepare the selected areas for the farmed salmon, all the native fish had been poisoned, so the failed fish farms, in addition to costing millions of dollars, removed dozens of wild salmon runs.

Hatcheries continue to release many millions of salmon every year, and the salmon returns have continued to decline. In 1992, a study conducted by Terry DiVietti, a Central Washington University psychology professor, provided an explanation. When fish are raised in a hatchery, they learn far less than the tiny fingerlings that grow up in the riverbed. Hatchery fish in concrete tanks have no threats in their life; they flock to the surface for food, and swim in packs the rest of the time. By contrast, fish that grow up in a stream learn to survive dozens of predators, and will hide, dart out, and zip back under cover to eat. They swim alone. They know how to use rocks and logs to conserve energy, having honed their swimming technique in the variable, complex flows of a stream. Hatchery fish learn none of these lessons in their tank. Releasing hatchery fish to the wild may be the piscine equivalent of sending a well-fed adolescent who has watched a lot of television into the woods to survive on his wits.

The other problem with hatchery fish is genetic. In the last few decades, it has been found that stream fish are much more genetically dynamic than anyone had ever imagined. There once were roughly 1,000 breeding stocks of salmon, of which 106 are now extinct and 314 are threatened or endangered. A human being offers up a few genetic carriers to the world; a salmon offers up thousands of genetically unique offspring, and the few that make it through to adulthood are likely among the very best of the lot. Therefore, salmon species adapt relatively quickly to local conditions. Genetically speaking, each stretch of river is home to its own strain, and every adult that returns is the pick of a very large litter. Hatcheries provide the protection that salmon are designed to do without, so hatchery-raised fish are genetically unculled. Releasing millions of hatchery fish plays havoc with the wild salmon, for both compete for the same food sources. Hatchery fish are much less likely than wild salmon to survive to adulthood, but they do apply pressure on the surviving wild stocks.

Thanks to dams, habitat degradation, overfishing, and hatcheries, the Pacific salmon has now disappeared from over 40 percent of its former range in Washington, Oregon, Idaho, and Montana. In 1994, despite a billion-dollar salmon recovery program, only about two and a half million salmon returned to spawn in the Columbia and its tributaries (two million hatchery fish, half a million wild), and the numbers continue to fall.

When anadromous fish swim up rivers to spawn, they provide an input of nutrients gathered at sea which are enjoyed by the entire food chain. Anadromous fish are consumed by carnivores, like wolverines and eagles, and omnivores, like bears and people; and the plants benefit as well, because fish concentrate phosphorus, one of the critical nutrients for plant growth, in their bones. The fish-laden feces of larger animals and the offal and bones left on land once delivered a nutrient element that present-day inland soils are often short of. When the spawning runs of fish declined, there was less flesh to sustain the meat eaters, and the green plants at the bottom of the food chain had less phosphorus to grow on, so the deer and elk had fewer new shoots to nibble on. No catastrophe, mind you, but it's harder to be fat and happy when the fish don't run.

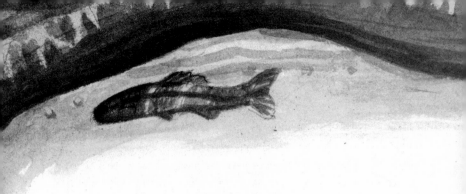

EIGHT

❧

MUSSELS, GATORS, AND THE CORPS

About one-third of the world's freshwater mussel species—some three hundred of them—are found in the waterways of the United States, and their names are a colloquial smorgasbord: purple pimplebacks, three-horn wartybacks, fat pocketbooks and spectaclecases, rough fatmuckets, pink heelsplitters, turgid blossoms, and Waccamaw spikes. Measuring up to 10 inches long, freshwater mussels burrow into the river bottom. Their shells lie slightly ajar when they feed, and their mantles—frills of golden flesh—reach up into the water to filter out the impurities for food, cleaning the flow.

Most species of mussels are slow-growing and long-lived: many take a half dozen years to reach maturity and live for periods varying from fifteen to fifty years. Some genera prefer swiftly flowing streams with gravel bottoms; other are found in muddy lake beds or sluggish, silty rivers. In the United States in particular, freshwater mussels exploited their niche with great success, and dense beds of mussels once covered the bottoms of streams, lakes, and rivers, clarifying the water.

Mussels are not mobile as adults, but when they are young they live for a while on the fins or gills of fish. Larval mussels, called glochidia, are discharged from the female mussel's marsupial-like pouch into open water. The larvae need to attach themselves to a suitable fish quickly, so mussels have evolved a number of strategies to lure fish into serving as obligate hosts. Some wave attractive body parts to draw fish near, while in other species the larvae themselves are easily mistaken for worms or minnows. Once safely encysted on a fish, the glochidia grow and eventually undergo metamorphosis into adults; that done, they drop off the fish and begin an independent existence in burrows at the bottom of the waterways.

This arrangement allowed the mussels to use the motive power of fish to spread throughout entire watersheds, leaving no likely habitat uncolonized. Inevitably, certain mussels evolved a dependence on specific host fish. The yellow sand shell, for example—one of the largest and most beautiful of the freshwater mussels in the Mississippi drainage system, renowned for producing large, lustrous yellow-tinged pearls—confines its attention to the gar, a once-plentiful species that became scarce when commercial fishermen targeted it as a destroyer of commercially valuable fish. (And not without reason; it is estimated that a single gar will eat 10 pounds of flesh for every one of its own, and then some.) As its numbers dwindled, so did the population of yellow sand shells.

Some mussels adapted to depend on specific fish, and some fish adapted to exploit the mussels. The sheepshead is a huge stout-jawed fish with round plates in its jaws and throat, perfect for crushing mussels and separating the flesh from the shell. When mussels are shattered this way, thousands of glochidia are released, and they encyst themselves on the sheepshead that has just eaten their parent.

Freshwater mussels, like their cousins the saltwater oysters, make pearls. The various species of mussels in American waterways, with different shades of nacre lining their shells, produce pearls in a wide range of colors, from white to faint shades of ivory, pink, yellow, and salmon, on through purple, copper, red, bronze, and steely blue. The freshwater pearls most familiar today are the grayish-white Japanese pearls cultured in the mussels of Lake Biwa, in central Honshu. Lumpish and baroque, they barely suggest the unsurpassing loveliness of the finest freshwater pearls. Round and with luster that makes them seem almost translucent, the best of the freshwater mussel pearls are said to be more luminous than the finest pearls from saltwater oysters.

The North American Indians gathered freshwater mussels for food, pearls, and wampum, leaving small hills of shells on the shores near prolific mussel beds. With the removal of the Indians, the freshwater mussels multiplied undisturbed, and the waterways teemed with them. The mussels and their pearls had a long period of uninterrupted growth.

Although mussels are not physically active, when their shell is ajar they will clamp down so hard on anything that touches their mantle that they can be pulled right up out of the riverbed on the end of a stick. In 1857, a carpenter who was randomly killing mussels found an exquisite pink pearl of 93 grains in a mussel pulled out of Notch Brook, near Paterson, New Jersey. (A standard measure for pearls, a grain equals 64.799 milligrams. There are 3.086 grains in one carat, which is 200 milligrams.) Nearly perfectly round and over a half inch across, the carpenter's Brook find was soon named the Queen Pearl and bought for $1,500 by Tiffany & Company, which then sold it at a two-thirds markup to Empress Eugénie of France. Fortune hunters immediately descended on Notch Brook and the surrounding waterways. The furor abated in a few seasons, after about $50,000 in pearls had been found and nearly every mussel within miles had been opened, but pearl fever broke out about every ten years after that, for the remainder of the century.

According to George Frederick Kunz's *Gems and Precious Stones of North America* (1892), farmers often hunted for pearls in their spare time. In the late 1800s, a single pearl from Montpelier, Vermont, brought $300 (the price of nearly 15 ounces of gold), and most of the streams in the state are thought to have lost their mussels in the excitement that followed. A pink pearl from Murfreesboro, Tennessee, sold for $150, single pearls from Texas streams sold for up to $250, and ten thousand people were pearling in Arkansas in the last decades of the 1800s.

In the 1870s, two anthropology professors excavating an Indian mound in Ohio's Little Miami Valley found a cache of about sixty thousand freshwater pearls, drilled and undrilled— many of them spherical, and varying in size from $\frac{1}{10}$ to $\frac{1}{2}$ inch in diameter. The pearls were decayed and of no commercial value, but in the wake of their discovery many Ohio streams were carefully raked and scraped. In 1878, the mussels of Waynesville, Ohio, gave up about $3,000 in pearls.

In the summer of 1889, a number of dark pink, purplish red, copper red, and metallic green pearls were found in the streams of southwest Wisconsin, and three months later $10,000 worth of Wisconsin pearls had been sold in New York City. A set of ninety-three Wisconsin pearls fetched £11,700 in London in 1890, while a necklace of thirty-eight freshwater pearls from the United States sold in 1904 for 500,000 francs in Paris. After eight years, Wisconsin streambeds had yielded about $300,000 in pearls, and the mussels in many of Wisconsin's waterways were nearly exterminated.

The Mississippi was once home to many of the finest pearl-producing mussels—as well as nearly all the species of mussels whose shells could be used for pearl buttons. It was the pearl-button business that nearly did in the Mississippi mussel beds. In 1891, a German named Johannes Boepple pioneered the shell-button industry with a small pearl-button factory on the banks of the Mississippi in Muscatine, Iowa. Within three years, mussel collectors had harvested nearly 10,000 tons of shells from a section of river near Muscatine about 1½ miles long and 300 yards wide, and when that bed was depleted the gatherers extended their quest. By 1897, 167 miles of the river were being harvested.

With ample raw materials and a ready market, pearl-button factories sprang up throughout the region. From the start, these factories were supported by itinerant shellers, who followed the mussel frontier in shantyboats. Mussel gatherers lived and worked in makeshift riverside camps, and temporary cutting plants were set up until the beds gave out. Although the shell factories accepted only certain species of shells, pearls provided roughly a third of the sheller's income, so the shellers would steam open their entire catch regardless of species, in large, flat pans. After removing the pearls, they would discard the meat and wash, sort, and market the shells. When a bed was first harvested, mature mussels with large shells were taken. After a few years, the average size decreased, and eventually only very small shells were left. Gatherers had to collect two or three shells to get the value of one, and as the supply fell the price rose, stimulating the shellers. Finally, when an area had been thoroughly depleted and the shellers moved on, the local population continued to work the depleted mussel beds periodically for extra cash, so the beds were unable to reestablish themselves.

At the turn of the century, there were sixty button factories along the upper Mississippi in Iowa, Illinois, Missouri, and Wisconsin, employing some two thousand workers. Most of these shops were "saw works," where rough blanks were cut and then shipped to New York and other Eastern cities for processing into completed buttons. In the winter of 1898–99, a shell buyer in Le Claire, Iowa, shipped 1,000 tons of shells to New York, representing a harvest of perhaps ten million mussels. The adoption of automatic machinery in 1901 accelerated the pace. The annual output of completed buttons shot up from 1.6 billion in 1904 to an all-time high of 5.7 billion buttons in 1916, by which time shelling had spread throughout a large part of the Mississippi Basin and into a few streams that drained into the Great Lakes and the Gulf of Mexico.

The mussels were gathered commercially by a tool called the crowfoot, made of an iron bar with up to a dozen four-pronged hooks attached to it. The crowfoot was dragged along the riverbed by a boat, and the mussels would snap their shells shut on the hooks. Dragging a bar bristling with iron hooks repeatedly along the riverbed is not good for growing mussels, and the shellers were persistent. In 1907, the U.S. Bureau of Fisheries guaranteed the mussel supply and funded a program of artificial propagation. The commercial propagation of mussels began in 1912, concentrating on the upper Mississippi, the Wabash in Indiana, the Ohio River, and the White and Black Rivers in Arkansas. Mussel glochidia were put into tanks with host fish, which were released to the rivers with the mussels safely encysted. This program dovetailed nicely with another bureau project—the fish rescue program, begun in the early 1900s. Each spring, the Mississippi would overflow its banks to spread into adjacent forests. River fish would spawn in these lowlands, and when the water receded millions of fingerlings were stranded in the slowly evaporating backwaters. Believing that fingerlings returned to the river would reach maturity and be caught by commercial fishermen, thus enriching the markets of the nation, the bureau hired crews to bring these landlocked fish back to the Mississippi. Before the rescued fish were released, many became encysted with glochidia, so the fish rescue and mussel propagation programs reduced overhead costs for each other.

With the mussel propagation program, the fisheries bureau hoped to save an economically important, highly competitive industry from self-destruction, but they did little to stop the use of the pernicious crowfoot; not until 1933, by which time the industry was moribund, did they state that if its use was allowed to continue the mussel beds would not recover. Between 1914 and 1920, the Mississippi produced 35,000 tons of shells a year and supplied about two-thirds of the world's buttons. At the height of the boom, the button business employed about twenty thousand people in the fisheries and factories, manufacturing $12.5 million of buttons a year. But it was the beginning of the end.

In 1930, the Bureau of Fisheries abandoned its mussel-stocking program, because it was an utter failure, and urged the five states along the upper Mississippi to repeal their mussel-protection laws—the dying industry, instead of being hobbled, should be given free rein. "It appears that the mussel fishery of the Mississippi River is doomed to economic exhaustion," wrote the commissioner of fisheries in 1932, and the bureau's official stance was that all the remaining mussels should be harvested.

Harvesting had caused local disruptions in the mussel beds, but when the waterways were engineered for transportation and flood control, the fish populations fared as poorly as the mussels. At one time, the Mississippi River supported abundant populations of the continent's largest freshwater fish: the white sturgeon (*Acipenser transmontanus*), a living relic covered with bony plates, topped out at 1,800 pounds; the paddlefish, a long-snouted, weird-looking fish, the sole survivor in this hemisphere of an ancient family (there is only one living close relative, in China) routinely weighed 100 to 200 pounds. Sturgeon and paddlefish were bottom feeders, while the gar pikes and bowfin, two oversized relics of ancient, purely American families, were voracious predators. Catfish, found everywhere but in swift mountain streams, grew to great size on the Mississippi riverbed: mud cats, channel cats, blue cats, and Mississippi cats often weighing over 100 pounds were caught, as were buffalo fish, blue suckers, and the sheepshead that fed on the mussels.

The lower Mississippi also supported dense populations of alligators, until the mid-nineteenth century. Another keystone species, alligators increase the complexity of swamps and marshes by building large mounds as nests and digging gator holes to overwinter in. In 1826, John James Audubon wrote that on the banks of the Red River as it neared the Mississippi, alligators "could be seen along the shores by the hundreds, or on immense rafts of floating or stranded timber, the smaller on the backs of the larger, all groaning and bellowing like thousands of irritated bulls about to fight." If an alligator hollowed out the same hole for several years in a row, the hollow would become a small pond. During droughts, gator holes sometimes provided the only open-water habitat available to amphibians, fish, and the predators that depended on them.

Alligators also maintain open-water habitats by tearing out the aquatic and emergent vegetation. In the spring, a female alligator will scoop up mud in her jaws, mix it with mouthfuls of vegetation, and build a mound 5 to 7 feet across at the base and roughly 3 feet high, in a shaded area near the water's edge. She lays between twenty and seventy eggs in the hollow on top of the mound, covers up the eggs, and tamps down the top of the mound with her feet to smooth it out. She then guards the mound faithfully, until late summer or early fall, when she's alerted by the peeping sounds of baby alligators and carefully digs up the eggs to free her children, 8-inch-long miniature alligators emerging from 3-inch-long eggs. Carrying them gently in her mouth, she takes the hatchlings to the water and watches over them until they are able to manage the swamp by themselves.

Thousands of alligator mounds once served as dry spots in the swamps along the Mississippi, providing nesting areas for turtles and colonization areas for plants, which the birds could nest in. Trees and shrubs root on old alligator mounds, so the Mississippi alligators often initiated the transformation of a marsh into a lowland forest. Between flooding and alligators, the entire length of the Mississippi River was once fringed by a complex edge of swamps and backwaters. Pockets of sedges, pond weeds, and millets fed huge assemblages of waterfowl. From the wild rice marshes of Minnesota to the coastal wetlands of the Delta, the Mississippi flyway, fed by the river's riparian edge, was an aerial highway. It was used by tens of millions of migratory ducks, which followed the seasons between summer nesting grounds in northern Canada and wintering areas near the Delta, while many more millions of birds flew it en route to Central and South America.

Floods are an annual occurrence on the Mississippi, and the broad hardwood forests of the lower Mississippi Valley were inundated seasonally. Floods expanded the habitat for many aquatic species, creating prime acreage for foraging, spawning, and a safe nursery for those vulnerable first weeks after hatching. Many fish made regular migrations to use flooded forests to spawn, and annual floods provided an enormous nutrient pulse to the land.

The first levees on the Mississippi were built in 1699 by individual landowners to confine the floodwaters, but work on the river was generally restricted to removing snags, bars, and caving banks, until the waterways were used more intensively for transportation. The first steamboat was built in Pittsburgh in 1811; eight years later, there were 191 steamboats plying the river; by 1833, more than 1,200 cargoes were unloaded at New Orleans. The cost of shipping freight between Cincinnati and New Orleans fell from 9¢ a pound in 1814 to less than half a cent in 1828, bringing an era of unprecedented prosperity to the river. Steamboats contributed to the widespread clearing of the riverside forests, for most of these vessels were wood-fired until the 1880s, when coal took over. The riparian zone was always the first to be cut. This cutting, combined with agricultural clearing, contributed to bank instability, while increased sediment loads and runoff from agricultural land increased bank caving as well.

In the early 1800s, much of the Midwest, particularly areas of Iowa and central Illinois, was swampy and poorly drained. This swampland was infested with malarial swarms of mosquitoes, but when ditches and channels were dug to shunt off the water, the marshes became rich bottomland fields. The work was done with horse-drawn slip scrapers, gangs of men with shovels, or enormous ditching plows, the largest of which required sixty-eight oxen driven by eight bullpunchers. Later, the dredging was done by huge floating machines, which were built on site, and most of the channels required regular dredging. The dredging of waterways containing mussel beds killed mussels directly, while mussel beds downstream were clogged with silt.

Channeling has drained more than 200,000 square miles of wetlands across the country, acre by acre and farmer by farmer. The states bordering the upper Mississippi, the Ohio, and the Missouri Rivers have lost over 80 percent of their wetlands. What the beavers built and left behind, the farmers ditched and drained. The land was channeled, and the rivers were channelized as well—not just to drain land for agriculture but also to improve water transportation and to speed the water away to reduce local flooding (which increased the flood pulse downstream). Swamp reclamation and navigational improvements caused rivers to flow more rapidly to the sea, river levels fluctuated more than before, and both of these changes were unfavorable to most fish populations. Moreover, artificial drainage increased the efficiency with which the available moisture was collected and carried to the waterways, so deepening a river channel or constructing new drainage works eventually lowered the local water table.

≈

Much of the channelization was carried out by the Civil Works Branch of the Army Corps of Engineers. The Corps was first formed in 1802 and undertook the improvement of inland navigation in 1824. As the country grew, so grew the Corps. Soon it was not only digging ditches and canals for drainage and transportation but building dams to manage floods, irrigate land, and generate hydroelectric power to sell. But its primary goal was to turn every river into a highway for barges, and it has done an excellent job. The Corps currently maintains more than 19,000 miles of improved inland and intracoastal water-ways, and also local flood-control projects incorporating over 350 dams (which sold nearly 70 billion kwh of electricity in 1992), along with more than 9,000 miles of levees and flood-walls and 7,500 miles of improved channels; while the Soil Conservation Service has channelized an additional 8,000 miles of rivers. The larger projects are part of river basin planning, but the smaller projects have often been initiated by people that hope to profit from them. If, for example, a rancher grazed cat-tle on a floodplain that was underwater half the year, he might ask the local congressman to arrange a feasibility study; if the engineering surveys and cost/benefit ratio looked practical, contracts were awarded and the dirt began to fly.

Even as late as the 1960s, the natural water cycle was so far removed from the minds of the Army Corps of Engineers that it considered the riparian zone—the ecotone between water and land—an unacceptable waste of water. The plants that live along a stream and suck water from the saturated zone are col-lectively termed phreatophytes (from the Greek for ground-watered plants). In a particularly wretched episode of engineer-ing run amok, the Corps started stripping the phreatophytes from the riverside in arid regions to conserve the streamflow. Along the Rio Grande River in Texas, 95 percent of the whitewing dove habitat was removed when the riparian zone was razed, and four million doves disappeared. Along the Gila River in Arizona, the lush riverside habitat supported doves, quail, pheasant, peccaries, deer, and even a few bears, while the trees cooled the water and supported generous populations of smallmouth bass and other fish. By the time the Corps bull-dozers were through, the river had been transformed into an overheated channel that supported neither fish nor game.

The purpose of channeling wetlands and rivers was twofold: to dry up the land and to improve water transportation, and on both counts the Corps was remarkably successful. Wetlands became farmland, the river bottoms were deepened, and the river biota changed as well. Channelization and dredging removed the pools and riffles from the streambed, simplifying the habitat. As a rule, when this happens the larger fish disap-pear and tiny minnows, shiners, and darters remain. In a 1960s study of the Tippah River, the Mississippi Game and Fish Commission found before channelization a total standing crop of 877 fish per acre, weighing a total of 241 pounds. After chan-nelization, 99 percent of the fish were darters, shiners, and min-nows. The total standing crop had nearly doubled—to 1,498 fish per acre—but the total weight per acre was only 5 pounds. In studies across the country, stream productivity was seen to crash when the streams were channeled.

In 1879, Congress established the Mississippi River Commis-sion to improve the river for navigation. To cope with the insta-bility of the river's banks, a large-scale system of levees was ini-tiated in 1884 that restricted the normal migration of the channel and prevented the creation of oxbow lakes. To improve navigation, the Corps set out to construct a channel 9 feet deep and 250 feet wide at low water between Cairo, Illinois, and the Gulf of Mexico. Dredging began in 1896 and sent a steady flow of sediment downstream, disrupting the riverbottom commu-nity. In 1913, the first dam was built on the upper Mississippi, at Keokuk, Iowa. Extolled by a writer in *Iowa Magazine* as "the most colossal engineering feat ever attempted, not only rival-ing, but actually surpassing, the ancient pyramids and the sphinx on the Nile," the 53-foot-high Keokuk Dam backed up the river for nearly 65 miles. By the end of the 1930s, twenty-six locks and dams had transformed the upper Mississippi into a series of lakes.

After the Keokuk Dam was built, the U.S. Bureau of Fisheries kept a close watch on the riverine life. The commercial fish harvest from the Keokuk impoundment peaked at 900 tons in 1917, and began a long downward slide that reached bottom in the 1930s, when fishermen reported their take as 120 tons. By then, the mussels were all but gone, too. In Lake Pepin, along the Minnesota-Wisconsin boundary—one of the bureau's centers of artificial propagation—the mussel take fell from 3,000 tons of shells in 1914 to 200 tons in 1919, when sections of the lake were closed to allow the mussel beds to recover.

Meanwhile, domestic sewage, industrial waste, and eroded silt were entering the upper river in ever increasing amounts. Minneapolis and St. Paul added sewage, while deforestation, plowing, and the draining of wetlands increased the silt load in the water. Sewage combines with fine silt and is carried farther downstream than it is in clear water, so local waste disposal suddenly had a regional impact. When a dam is built, the sewage-laden silt that would have washed downstream settles to the bottom of the reservoir, where it breaks down slowly and severely depletes the oxygen content of the water.

In 1927, the floodwaters of the Mississippi topped the embankments, and more than 26,000 square miles of land in Mississippi, Louisiana, Arkansas, Missouri, Tennessee, and Kentucky were inundated. Entire towns were underwater; roads, rail, and telephone lines were out of commission; and 214 people were drowned. As a result of this flood, engineers took a closer look at the river. The Flood Control Act of 1936 brought about the modification of an estimated 16,000 miles of the Mississippi's main channel and its tributaries. The main channel was widened from 250 to 300 feet, and further dredging deepened it to 12 feet—an ongoing effort that has ultimately extended from the Gulf to Minneapolis–St. Paul. Between

1932 and 1955, the lower river was shortened a total of 152 miles by means of cutoffs dug through its meanders, and a further 54-mile shortening was planned. These cutoffs were constructed to reduce erosion and make the river's energy available to transport sediment, and they had a dramatic effect on riverine life. To keep the river in its new banks, the new channels had to be continually dredged for more than ten years, and between 1932 and 1955 over 1,300 million cubic meters of sediment were sent down the river. Finally, dozens of dikes and miles of revetments were constructed between 1945 and 1970, to help nip meanders in the bud.

A natural river writhes like a snake, throwing off oxbow lakes as meander loops slip downstream. But the flow in the lower Mississippi River has been transformed from a broad band of shifting riverbeds to a single, narrow, deep, swift channel. Walled in by 1,870 miles of main-stem levees, the Mississippi has been isolated from the surrounding floodplain, the riparian zone replaced by mattresses of concrete blocks. Fish that once relied on their annual visit to flooded hardwood forests for food, spawning, and nursery areas now have fewer such sanctuaries. The levees prevent the water from spreading out and dropping its silt, so the organic matter that once replenished the terrestrial ecosystem is now flushed out to the ocean, where it settles onto the continental shelf. The absence of silt results in coastal recession and subsidence and a shift in nutrient cycling: the nutrient elements that once fueled the estuaries are now deposited in deep water.

The Mississippi, dammed, channeled, and leveed, doesn't move much anymore. The rafts of logs and the alligators are confined to remote backwaters, the great fish that the river was once renowned for have slipped into legend, and the indigenous mussels, which once filtered the impurities from the water, are just about gone.

The upper Mississippi and the Missouri were once characterized by high turbidity, wide seasonal fluctuation in flow, and a wide, braided channel that constantly changed location. Bank erosion was continuous, and the channels shifted every flood season. Today there are twenty-six dams on the Mississippi and sixty on the Missouri. The Mississippi is a highway for barges, constricted to a dredged, leveed channel. Freshwater mussels are now the most widely endangered family of organisms in the United States, and malacologists (those who study mussels) say that the mussels are poised on the brink of a major extinction. One in every ten mussel species is thought to be extinct already, and two of every three of the remaining species are threatened. And when the great freshwater mussel beds disappeared, the water flowed unfiltered to the sea.

NINE

~

AQUEDUCTS AND TOILET BOWLS

One of the earliest problems of human civilization was keeping feces out of the water supply. Almost without exception, cities were sited near lakes, streams, or springs, which supplied the citizenry with fresh water. As the population grew, runoff and daily dumping polluted the surface water, while waste from outhouses, cesspools, and stables seeped down to the groundwater. Before long, the lake or river would be undrinkable and the wells and springs tainted. Feces and water are both vectors for the transmission of many diseases, and cities are dangerous places to live unless clean water can be piped in from remote locations and wastewater can be piped out in sewers.

The technical problems of water supply and sewer systems were solved early. By 2500 B.C., the Sumer and Indus Valleys had huge irrigation works, and the cities that grew from those well-watered fields were fully plumbed as well. Aqueducts brought fresh water from the mountains into the city to supply public fountains, baths, and latrines, while sewers shunted wastewater out of the city and back into the waterways. Some private Sumerian homes had water-flushed latrines that connected to the sewer system, a convenience that didn't become commonplace until four millennia had passed.

Roman aqueducts and sewers have long been admired as one of the marvels of the ancient world. In 97 A.D., the former governor of Roman Britain, Sextus Julius Frontinus, was appointed superintendent of Rome's aqueducts, and his *De Aquis Urbis Romae* is a comprehensive tour of the triumphs of Roman waterworks. Instead of drinking from the polluted Tiber or tainted local wells, the Roman population drank water brought to the city from the forest streams of the Apennines, in aqueducts built of local stone and lined with cement. By the time Frontinus wrote, the Romans had built 255 miles of aqueducts carried on arched bridges across valleys, tunneled through rises, and supported on structures of solid masonry. Within Rome, water was distributed through lead pipes and hollowed-out logs to public baths and fountains, to 144 public latrines (some of which had 20 seats to a side), to shops, and to dozens of private homes. Water use per capita was about 300 gallons a day, roughly three times what it is now in the United States.

After the Roman Empire fell, Europe is said to have gone

unwashed for a thousand years. With the ascendancy of Christianity, the wealth that the Romans had diverted into the construction of aqueducts and sewers was spent on cathedrals and their bureaucracy, for the Church was concerned with loftier matters than its flock's temporal existence. Ironically, monasteries were the only places where plumbing was regularly found in the Middle Ages: many had piped water systems and latrines connected to sewers. City wells were invariably fouled and the streets were even fouler, but people tithed, prayed, and bought their drinking water from carts which hauled it from springs outside the city limits—and they washed only rarely, with water from tainted wells.

Which became progressively more contaminated, for Christians preferred to be buried in the sanctified ground of the churchyard, creating layer upon layer of putrefying parishioners, all leaching down to the groundwater. Over time, urban churches laid hundreds of thousands of believers to rest in a small area. The Scottish journalist Basil Hall writes in 1820 of churchyards ankle deep with "rank and offensive mould, mixed with broken bones and fragments of coffins," while the water of city wells near churches grew redolent with putrescine, cadaverine, and other pungent organic compounds associated with decomposition.

At the time of Independence, Boston, New York, and Philadelphia were drawing their water almost exclusively from wells, cisterns, and springs. As in the Old World, Americans buried their dead within city limits (the churchyard of New York's Trinity Church, for example, held a hundred and sixty thousand graves by 1830). The contents of cesspools and outhouses leached into the groundwater, and city wells went from bad to worse. By 1809 the water from New York City's public pumps was so foul that horses balked at drinking it; by 1835, one-quarter of Boston's 2,767 wells yielded water too putrid to drink. "The smell in many cases is extremely offensive, and I should think it probable that they have an injurious effect on the water of wells contiguous," an engineer's assistant noted that year. But odor and taste were relatively unimportant in an era when no one washed and everyone drank ale: it was disease, rather than malodorous wells, that prompted American cities to build reservoirs and aqueducts.

In this country's first century, American cities, like those of the Old World, were host to repeated epidemics, which killed more and more people as cities increased in size. No one knew how disease was transmitted, but there were two leading hypotheses. The miasmatics subscribed to the theory first put forth by Hippocrates more than two thousand years earlier: disease was caused by miasma, memorably characterized by one observer of the great 1665 plague of London as "a most subtle, peculiar, insinuating, venomous, deleterious exhalation arising from the maturation of the ferment of the feces of the earth." Contagionists believed that disease was passed from person to person, and that epidemics could be prevented by quarantining the city from foreign ships, halting trade. The miasmatics prescribed washing away the city's accumulated filth with plenty of clean water, a task of Augean proportions.

Considering what was clogging the streets, the miasmatics may well have been on the right track. Garbage was dumped in the alleys to be eaten by bands of feral pigs. Chamber pots (said to have been invented by the Sybarites, who were too lazy to go outside) were emptied out the door or into household cesspools. And when horsepower was produced by hay rather than fossil fuels, there was an enormous amount of manure to contend with. In 1780, when the population of Paris was about 600,000, city street sweepers removed 270,000 cubic meters of manure, cesspool collectors took a tenth of that, and the whole load was conveniently dumped into the Seine. The cesspool collectors would have accounted for only about a third of the feces produced annually by a city of

600,000, so the rest must have gone into the street. If Philadelphia was like Paris but one-tenth the size, the street sweepers collected about 30,000 tons of droppings annually, of which perhaps a fifth was human waste, while cesspool collectors managed 3,000 additional tons.

Yellow fever, which had entered the New World in 1647 with African slaves brought to Barbados, struck Philadelphia with a vengeance in 1793. Almost four thousand people died, and twenty-three thousand fled, effectively closing down the nation's capital and leading port for months. The disease skipped up and down the Eastern seaboard for the next few years, and in 1798 another thirty-eight hundred Philadelphians died and forty thousand people (about three-quarters of the population) left. The city's leading medical light, Dr. Benjamin Rush—best known for his *Medical Inquiries and Observations upon the Diseases of the Mind* (1812), one of the first modern attempts to explain mental illness—was a miasmatic. He proclaimed that yellow fever was caused by "the putrid exhalations from the gutters, streets, ponds and marshy grounds in the neighborhood of the city," and that the cure was to cleanse the streets with fresh water piped from the Schuylkill River. Given the alternative—closing the harbor—most of the wealthy Philadelphians heartily endorsed his scheme.

Scholars were classically educated in those days—they could read Frontinus in the original—and were familiar with the Roman method of providing a city with cheap and plentiful water. But there were relatively few civil engineers in the United States at that time (and no engineering schools), so the city of Philadelphia contracted with an English civil engineer named Benjamin Latrobe to set up their water supply system. (In her interesting and artlessly named book *Bathrooms*, the interior designer Mary Gilliatt credits Latrobe with constructing the first bathroom in the United States to have a bathtub and sink as well as a toilet—in a Philadelphia house, in 1810.)

By 1837, the city had spent about half a million dollars building a dozen miles of aqueducts and four reservoirs with a combined capacity of 22 million gallons, and laying over 98 miles of iron distribution pipes within the city. Nearly fifteen hundred houses had running water, and the streets were cleared of filth. "Blessed as we are above every other city of the Union, and perhaps not excelled by any in the world in the cheap and abundant supply of pure and wholesome water we now enjoy," burbled the *Annual Report of the Watering Committee* for the year 1836.

As yellow fever catalyzed the construction of Philadelphia's first aqueducts and reservoirs, so cholera triggered New York's. Cholera entered the New World via Montreal in June 1832, popping up in town after town until it reached New York City a month later. By October, thirty-five hundred New Yorkers were dead and a hundred thousand had fled the city. Philadelphia, with its clean streets and plentiful supply of untainted water, lost only nine hundred lives. Work on New York's Croton Aqueduct system—which included the Croton Dam, four reservoirs, and about 50 miles of tunneled and bridged masonry aqueduct at a total cost of over $10 million—began shortly thereafter.

The earliest city water supply systems were built by private companies, but as cities increased in size the proposed aqueducts began to rival those of the Romans, and the costs became so high that water supply systems were built by city governments. The funds were typically raised by selling bonds, and recouped by charging an annual fee of $10 or so for each householder's hook-up. There was no lack of takers: in 1849, nine months after the Cochituate Aqueduct opened in Boston, there were 10,851 private subscribers and 750 fire hydrants.

Every new subscriber to the water supply system increased wastewater flows, and the need for municipal sewer systems became increasingly apparent. Sewers were a necessity even before a municipal water supply system was built, when the water supply was from city wells: cellars would fill up during rainstorms and the streets became quagmires if the water had nowhere to flow. In unsewered areas of a city, the streets were drained by digging depressions at critical intersections to collect the flow from the gutters. These drainage basins often became open cesspools, in which dead cats and dogs and garbage of all sorts piled up. In 1806, the British traveler John Melish wrote that New York City's drainage pools "accumulate such a collection of latent filth, that the steams of it are sometimes perceptible at two miles' distance."

In colonial times, the homeowners along a street would band together and dig their sewers collectively, laying a pipe directly to the nearest waterways. When kitchen sinks that could drain to a sewer replaced dry sinks, life was made much easier, so people went to great lengths to lay pipes. The expense was shared, and when latecomers connected to the sewer, they paid the builders for the privilege. The sewer network extended block by block, informally, until water supply systems increased the flow to the city and the jerry-built sewers began to back up.

When Boston officially became a city, in 1823, one of the first acts of its government was to take over ownership and control of the area's sewers. The original lines had been laid for household drainage, but the city needed to build storm drains to clean the streets and empty the drainage basins. In the next fifteen years, the city laid scores of miles of sewers large enough to be entered for cleaning and artfully crafted of brick and stone. The cost of construction and maintenance was charged to connecting homeowners based on their assessed valuation, but since the householders couldn't be disconnected, the bills were rarely paid. Between 1823 and 1838, the city spent $121,109.52 on sewers, and less than a quarter of the cost was collected from the homeowners.

Water supply and sewer systems allowed cities to clean the streets by washing the wastes that were clogging the streets into the nearest body of water. Sewers had not been laid with feces in mind, however; kitchen sinks were the household connection to the sewer, and Boston (for example) banned human wastes from its sewers until 1833. Popular acclaim for the flush toilet changed the central role of the sewers, the scent of the city, and the layout of residences as well.

There were no bathrooms before the advent of indoor plumbing: portable bathtubs were filled in the kitchen, washstands stood in each bedroom, and chamber pots hid under each bed. With a municipal water supply came bathrooms with fixed tubs that no longer had to be drained by hand, a plumbed bathroom sink rather than individual ewers, and toilets, which caught on fast once the sewer systems were built. Though the bowls didn't always clear and the flush was noisy, the toilet so outperformed the chamber pot that to use one was to desire one.

The father of the flush toilet is thought to be Sir John Harington, a godson of Queen Elizabeth I. Sir John described his contraption in *The Metamorphosis of Ajax* (a pun on "a jakes," or outhouse), a book he published in 1596. His toilet had all the essentials but one: the pipe left the bowl in a straight line, so the water closet smelled like the cesspool below. (Queen Elizabeth is said to have complained about the smell, at a time when people were relieving themselves almost anywhere.) That problem was remedied in 1775, when the British engineer Alexander Cummings patented a toilet with a siphon trap—a pocket of water in a lazy-S pipe—that kept the smell of the cesspool (or sewer) from wafting back into the bathroom.

As soon as American cities had dependable water supplies and a sewer system, all the fashionable hotels installed bathtubs, sinks, and toilets, and the wealthier citizens followed suit. Rather than squatting on chamber pots and hauling buckets of water from pumps or carts, people were turning on the tap and enjoying the pleasures of a 5-gallon flush commode. In 1860, there were 6,500 flush toilets and 100 miles of city sewers in Boston, and the hotels used simply stunning amounts of water: Boston's Tremont House used over 25,000 gallons a day, and the Parker House over 20,000. By 1885, Boston had 100,000 toilets, 226 miles of sewers, and literally thousands of miles of pipes bringing wastewater to municipal sewer mains 8 feet in diameter.

Families with bathrooms use hundreds more gallons of water a day than families without, and water consumption skyrocketed when bathrooms became commonplace. During the century in which Boston went from wells to flush toilets, daily per-capita water consumption jumped from perhaps a dozen to nearly 100 gallons a day—about the present rate. In the course of building water supply systems, a wholly new standard of living was born.

Meanwhile, advances in microscope technology had revealed the presence of "animaliculae" in perfectly clean water, an alarming discovery. A pamphlet printed in 1845, shortly after the discovery of microorganisms in Boston's drinking water, showed an old Irishman peering into a microscope. "Well, did anybody ever see the likes of this!" he exclaimed to the crowd around him. "They call whiskey 'a droph o' the craiture.' But here's water that's a droph of a thousand craitures!"

Before these unsettling revelations, water had been judged on the basis of its scent, taste, and clarity, and there was a nearly universal belief that water that didn't offend the senses was above suspicion. The discovery that seemingly limpid water was actually teeming with tiny creatures was highly disturbing; the New York diarist George Templeton Strong noted on August 1, 1842, that people had "dreadful apprehensions of breeding bull-frogs inwardly."

In spite of public interest and functional microscopes, the

cause of disease wasn't fully formulated until 1880, when Louis Pasteur published *On the Extension of the Germ Theory to the Etiology of Certain Common Diseases* and replaced the concept of a nebulous malificent miasma with the theory that specific air- or waterborne organisms caused specific diseases. In the next five years, medical researchers identified the microorganisms responsible for gonorrhea, malaria, typhoid, tuberculosis, diphtheria, cholera, and tetanus.

The discovery that would have perhaps the greatest impact on the nation's municipal water systems came in 1884, when the German pediatrician Theodor Escherich put the feces of a cholera patient under a microscope in search of the organism that causes cholera. Those were the days when the world of science was an unwritten book, for the organism Escherich isolated, *Escherichia coli*, is hard to miss: they make up one-fifth to one-third of the weight of an average feces. Sure enough, further investigations showed that *E. coli* are also present in the feces of healthy individuals. Humans produce about two hundred billion coliform cells a day, making the amount of *E. coli* in waterways an ideal indicator of the water's fecal contamination. The odds of seeing a typhoid bacillus in a sample of water were slim, but a high count of *E. coli* was proof enough that the water was potentially infectious.

Sewering the city's wastes into the nearest body of water improved the health standards of upstream communities at the expense of those downstream. Typhoid fever, carried by feces, had become a significant cause of mortality: in 1880, the death rate from typhoid fever in New York was about thirty-two out of a hundred thousand; forty-two in Boston; fifty-eight in Philadelphia; and fifty-nine in Baltimore. Lawrence, Massachusetts, which drew its water from the polluted Merrimack River, had a typhoid mortality rate that remained eighty or more per hundred thousand from 1887 to 1894.

When watertight household vessels were developed in pre-

historic times, water was clarified by storage, and also by filtration through porous receptacles or sand layers: in both cases,
sediment was left behind. After many more centuries, it was
found that filtering and sedimentation could be aided by the
addition of precipitants and flocculants, which make the solids
in the water clump together and sink: this discovery marked the
beginning of water treatment. Sanskrit medical lore and Egypt-
ian inscriptions from about 2000 B.C. are the earliest records of
what is now known as primary water treatment. Alum, the floc-
culant most widely used to clarify water today, was first mentioned by Pliny in 77 A.D.; he described the use of both lime
and alum for conditioning water. By the fifteenth century, the
preparation of purified alum had reached manufacturing
dimensions, and by 1767 the common people of England were
purifying water by adding two or three grains of alum to a
quart and allowing the impurities in the water to flocculate and
settle out, after which the supernatant was filtered.

In 1872, Poughkeepsie, which drank from the Hudson, built
a sand filter and became the first American city with a filtered
water supply. While the use of flocculation, sedimentation, and
filtration of individual household and small industrial water
supplies had been practiced from ancient times, the earliest use
of flocculation for treatment of municipal water supplies
occurred at Bolton, England, in 1881. From 1885 on, flocculation was widely used as a process preceding filtration.

Few American cities filtered their water supply in the 1890s;
whereas twenty million Europeans drank filtered water, only
about a hundred thousand Americans did. Just a few years after
E. coli was named and typhoid, cholera, and diarrhea were
identified as waterborne diseases, filtration systems were shown
to remove microorganisms from the drinking water. The result
was a flurry of construction. Communities that built filterbeds
had sharp declines in the rates of waterborne diseases, and the
results were widely publicized. In 1893, the city of Lawrence

built filters for its water supply, and the typhoid death rate fell from a high of 134 to between 20 and 30 people per 100,000. Innovations in mechanical filtration and sand filtration followed quickly, and dozens of articles and books plotted the death rate from waterborne disease against the type of city water supply.

In 1903, the waterworks pioneer Allen Hazen of Boston discovered that when a polluted city water supply was filtered, for every person saved from typhoid there were three saved from death by other causes, many of which had theretofore seemed unrelated to the water supply. This correlation between water supply and the mortality rate, known as the Hazen Theorem, brought filtration to the forefront of city agendas. No one had been very keen about drinking animaliculae in the first place, and the indication that they were associated with disease led most large American cities to build filtration systems for their water supply systems by 1910.

Chlorine was first used to purify water by William Cruikshank in England in 1800, though it wasn't applied to municipal systems until much later. Chlorination of public water supplies came to the United States in 1908, via the Chicago stockyards. The Union Stockyard Company had contracted with a private water supply company to provide the cattle with water from sewage-contaminated Bubbly Creek. The Bubbly Creek Filter Plant was built, and the water was filtered and cleaned by dosing it with copper sulfate, which kills algae; however, the cattle who drank the water didn't gain enough weight, so the stockyards started using city water again. The City of Chicago sued the Union Stockyard Company for poaching municipal water, and the Bubbly Creek Filter Plant then substituted chloride of lime (a source of chlorine) for copper sulfate; the cattle thrived, and the water was found to be cleaner than city water, in spite of its lowly origins. Later that year, Jersey City sued *its* water supply company for delivering tainted water; chlorine was added, and the courts ruled that the chlorinated water was legally and officially pure. Chlorination kills most microorganisms, and when chlorination is combined with filtration, the water supply becomes blameless.

The death rate from typhoid fever was the yardstick used by public health authorities to assess the public health of a community. As city water supplies began to be filtered and chlorinated, the average urban death rate from typhoid dropped from thirty-six per hundred thousand in 1900 to twenty per hundred thousand in 1910 to three per hundred thousand in 1935. Today, there are only a few dozen cases of typhoid fever a year in the United States.

Maps of municipal water supply and sewer systems show the vascularization of a city. Water is piped from pristine reservoirs or deep wells and diverted into hundreds of thousands of dwellings, where individuals turn on the tap and flush the toilets. The myriad of sewer lines join with many others to the sewer main, a great underground river with a daily flow that may be hundreds of millions of gallons, carrying the feces from millions of people to be discharged at a single point into the waterways. New York City's water supply system now includes more than 350 miles of aqueducts and 27 man-made lakes, which supply about 1.5 billion gallons of water a day—approximately the same as the flow of the upper Hudson or of the Colorado at the Arizona-Colorado line.

When organic matter is dumped into a stream, it is broken down to its constituent parts and eaten by microorganisms, which require oxygen to metabolize their food. Since the aquatic ecosystem is fueled by the organic matter in water, running water purifies itself, and the accepted wisdom among civil engineers and the public alike was that the waterways could be utilized for waste disposal almost without restriction. Water filtration and chlorination could protect the public from water-borne disease.

Industries that dumped their wastes into the water in the nineteenth and early twentieth centuries included beet-sugar processors, canneries, pulp and paper mills, textile mills, dairies, meat packers, and tanneries, most of which laid a private pipe directly to the nearest body of water rather than connecting to the city sewers. Most of this industrial waste was simply organic matter (with the exception of effluvia from the tanneries and mills, both of which used an array of chemicals to break down fibers), and as long as there was enough oxygen in the water-ways, a stream or river could assimilate them—providing that the life in the water was not overwhelmed and asphyxiated.

But by 1912 there was enough concern about water pollu-tion to warrant Congress's mandating a study of stream pollu-tion. The Public Health Service funded a group of sanitary engineers, chemists, biologists, bacteriologists, and doctors to find out how much waste a stream could absorb, how much oxygen was required in order to break down different types of waste, and the efficacy of various treatment methods for indus-trial waste. One result was the "oxygen-sag curve," which defined how much waste a stream could assimilate. The dis-solved oxygen of a stream could be measured, and the amount of oxygen needed to break down various wastes was character-ized at specific times and temperatures. The dissolved oxygen was like cash in a bank, continually replenished by continuing deposits of aeration. The wastes were checks, which could be drawn against it.

The so-called Streeter-Phelps model, named for the two san-itation engineers that developed it, became the standard method of estimating the effects of industrial and human pol-lution on streams, and in some cases the oxygen studies were combined with bacterial surveys to form measures of stream sanitation and the adequacy of dilution. Industrial wastes were described in terms of their population equivalents—that is, a unit of industrial waste was expressed in terms of the equivalent amount of wastes of a number of people. Unfortunately, oxygen limits are less relevant when inorganic or toxic components are added to industrial waste, a drawback that would not become apparent until the years following the Second World War.

TEN

❧

DOWN THE DRAIN,
UP THE STACK

A revolution in chemistry took place in the years before 1940, and the urgency of war and the stimulus of postwar reconstruction put this new knowledge into practice. The technological revolution that followed the Second World War changed industry, agriculture, transportation, and communications. New industrial processes and products created literally thousands of new chemicals, and industrial effluent became increasingly toxic. Waste heat from power plants was also disposed of in the waterways, reducing the oxygen content of the water. Urban populations began to swell, and so did the flows of sewage. Mining increased in scale, contributing its own pollutants to the waterways, and agriculture adopted chemical fertilizers, insecticides, and herbicides, many of which washed into streams and rivers. The effects of all these new pollutants on the already simplified aquatic ecosystem was disastrous.

Better living through chemistry was the ethos of the postwar era, and the production of synthetic organic chemicals went from less than 1 billion pounds in 1940 to over 160 billion pounds by 1970; today there are some seventy thousand widely used synthetic chemicals, and about a thousand new chemicals are introduced each year. Many of these newly created substances were spread throughout the environment, and some of them became notorious.

DDT—dichloro-diphenyl-trichloroethane—began to be marketed as an insecticide in 1944, the first of a family of chlorinated hydrocarbons that soon included 2,4-D (the herbicide Agent Orange), heptachlor, Dieldrin, Endrin, Aldrin, and Mirex, among others. A microscopic speck of these substances posed a substantial cancer risk; a pinch could kill you outright. When DDT first came on the market, its benefits were clear: it killed agricultural pests and thereby reduced the cost of food. Its drawbacks were less obvious.

Until DDT was released to the environment, natural processes were thought to dilute chemicals to the point where they virtually disappeared. It's a reasonable assumption, except that the biosphere tends to concentrate certain molecules. When DDT was swept into the waterways even in concentrations as small as 3 parts per trillion—that is, roughly a drop for every 3 million gallons—the life in water went to work. Zooplankton collected DDT at 40 parts per billion: the minnows that ate the zooplankton held 5 parts per million of DDT in their fat; and the fat of fish-eating birds, like eagles and pelicans, held concentrations of 25 parts per million. In a relatively short time, DDT could be found in organisms at 8 million times the level in the water. At those concentrations, it interfered with calcium deposition on eggshells, and birds high on the food chain began laying eggs encased in membranes instead of shells. As a result, brown pelicans, ospreys, bald eagles, and other fish-eating birds nearly disappeared.

PCBs—polychlorinated biphenyls—were first manufactured in 1929, and were used as plasticizers and to produce paints and resins. Wonderfully inert, the new substance was widely used in electrical insulators. PCBs were known to be toxic, but since insulators were meant to be permanent installations, why worry? By the time it was understood that PCBs bioaccumulate and insulators are ephemeral from the earth's point of view, seals were having pups with fused fins. PCBs, as it turned out, are teratogens as well as mutagens and carcinogens, deforming the offspring of mammals at sublethal concentrations.

Some of the chemicals that backfired, like DDT and PCBs, were useful products, manufactured and introduced into the environment on purpose. Other exotic organic chemicals and heavy metals were dumped into the streams and rivers as unwanted by-products of industrial processes. Mercury, for example, was used as a catalyst in the production of caustic soda and acetaldehyde. In the 1940s, about 6 ounces of mercury were discharged into the waterways for each ton of caustic soda produced. Mercury was also used in the manufacture of the plastics urethane and vinyl chloride. The electrical industry used tons of mercury in batteries, silent switches, high-intensity street lamps, photocopying machines, and fluorescent lights. In 1969, the annual consumption of mercury in the United States peaked at over 3,000 tons.

It was thought that the mercury discharged into waterways settled to the bottom sediments and remained there, inert. But anaerobic bacteria in the bottom muck convert mercury into an organic compound called methyl mercury, which is readily concentrated by living organisms, including people. In Minimata, Japan, the local chemical plant had begun making acetaldehyde—and discharging mercury into the ocean—in 1932. Mercury discharges soared in the early 1950s, when the factory became the sole Japanese producer of a plasticizer that used acetaldehyde in its manufacture. The mercury was duly concentrated by the aquatic foodchain. In April 1956, there was an outbreak of a noncontagious illness of the central nervous system that would later become known as Minimata disease: the cats in town began to die, and the pediatrics ward of the Minimata factory hospital was crowded with brain-damaged children. Local fish were suspected as the source of the poison, and two years later the Kumamoto Prefecture banned the selling of Minimata fish. Mercury was identified as the cause of the disease in 1959, and the factory reduced its mercury discharges but did not stop them altogether until 1968. In 1969, the factory owners were taken to court, and after four years of legal wrangling, they compensated more than three thousand victims of brain damage.

Dioxin, a highly toxic carcinogen that affects the immune and reproductive systems of mammals at very low doses, is an industrial by-product with a twist: the molecule is created by the processes themselves, a side effect no one expected. Dioxins, along with other similar toxic molecules, are created whenever organochlorinated molecules are produced or destroyed. Dioxins occur in the chlorine bleaching process used to make paper, and in the incineration of such materials as PVC plastics or wood treated with pentachlorophenol preservatives. It is impossible to ban dioxins without banning the industrial processes that create them.

Some of the new chemicals released to the environment have been found to interact with the body's receptors for estrogen, which appear to be not so good at recognizing what is a hormone and what is not a hormone. Nonoxylphenol polyethoxylates, used as surfactants in dishwashing liquids and toiletries, are estrogen mimics. Polycarbonate plastic, widely used in household containers and appliances, sheds bisphenol-A, an estrogen mimic so potent that 2 to 5 parts per billion are enough to induce increased hormone responses in reptiles, fish, shellfish, birds, and mammals. Endosulfan—one of the most widely used pesticides—acts like estrogen, and even the trace quantities left as residue on fruits and vegetables is enough to affect hormonal activity in many species, including human beings. DDT, PCBs, dioxins, atrazine, and some forty other compounds (many of them common household chemicals) have all been identified as hormone influences. Although these chemicals do not increase the levels of estrogen in an individual, the body perceives increased hormone levels. It is not clear how females are affected, but males of numerous animal species, from panthers to alligators, were found to have reduced sperm counts and smaller testes as a result of exposure to these chemicals, and some male fish develop genitals of both sexes. Worldwide, human sperm counts have dropped significantly since the 1940s, and a growing number of biologists believe that this may be due to environmental exposure to chemical estrogen mimics.

The chemical industry itself was responsible for most of the exotic new toxins in the waterways, but many other industries left their mark. Paper and pulp mills used massive amounts of sulfur, chlorine, sodium hydroxide, sodium carbonate, sodium sulfate, sodium chlorate, titanium dioxide, and aluminum sulfate, all of which were carried away in the wastewater. Textile production made heavy use of the waterways for waste disposal. When wool was carbonized, for example, strong acids were used to remove the cellulose impurities, followed by a soda-ash wash. Cotton fibers were coated with sizing to give them enough body to withstand the pressures exerted during weaving. Before 1960, the sizing was generally starch, and the desizing was done by enzymes, which had to be discarded into the waterways after each batch. Scouring of the fabric with hot alkaline detergents, caustic soda, and soda ash followed. And if the cotton was mercerized, sodium hydroxide and a sulfuric acid wash were added. Dying was a chemical nightmare, and the rivers running through mill towns were often rainbow banners of the daily changes in dye lots. Many of the synthetic fibers developed in the 1950s were water-repellent, so new sizings had to be developed. Instead of starch baths, polyvinyl alcohol, carboxymethyl cellulose, and polyacrylic acid baths were used to give the fibers body. Instead of enzyme desizing, sulfuric acid desizing was necessary for synthetic fibers like rayon and polyester, and the acid bath also ended up in the nearest stream, river, or lake. Molybdenum, chromium, vanadium, nickel, and other metals, many of them poisonous in trace amounts, were used in the manufacture of steel alloys. Pickling the steel, a tempering process, required significant amounts of cyanide. All these substances found their way into the waterways—and some interacted synergistically, becoming more toxic together than they were individually.

By the early 1960s, the sluggish Calumet River (to give just one example), which empties into Lake Michigan, was receiving a daily dose of about 100,000 pounds of oil, 35,000 pounds of ammonia, 3,500 pounds of phenols, and 3,000 pounds of cyanide from the dense industrial complex around Chicago, Gary, and Hammond.

The amount of organic oxygen-demanding waste added to the waterways increased as well. The food-processing industry used the waterways to dispose of the hair, offal, and blood from meat processing; the scum and whey from dairy products; the leavings from beet-sugar refining, brewing and distilling, canning and freezing. In the early 1960s, food processors in California's Central Valley dumped over a *billion* pounds of vegetable waste into the waterways each year during the canning and freezing season. Until the mid–1960s, the stockyards and packing plants of Omaha disposed of their wastes in the Missouri River, and a stretch of the river occasionally ran red with animal blood. Hair, entrails, and mats of congealed grease floated downstream and collected on islands and on the riverside.

City populations swelled after 1940, and the sewer systems

that had connected hundreds of thousands of toilets to a single main sewer now had millions of homes hooked in. More solids were discharged at a single point, and municipal wastewater treatment plants were built in many cities to prevent feces and toilet paper from fouling the beaches. Primary treatment plants removed much of the solids from the wastewater flow, but had little effect on the chemicals, including detergents.

Until 1946, people washed their clothes with soap, but as soon as synthetic detergents were sold they captured the market. Detergent molecules are long and asymmetrical: one end attracts dirt while the other end attracts water, which is why detergent lifts dirt from wet clothes. Detergent molecules come in two forms, straight and branched—a difference that affects sudsing but not cleansing ability. The suds of the straight molecules of detergent break down quickly, while the suds of the branched molecules break down slowly if at all. In the early 1960s, detergent manufacturers waged an advertising battle over who had the longest-lasting suds, and the waterways lost. Suds began to appear in streams, rivers, and lakes, and in many areas great mounds washed up on shore. At the foot of Niagara Falls, the piles of discolored detergent foam were 8 feet high.

Detergents also contained phosphate additives to soften the water and thereby improve the effectiveness of the detergent molecule. The environmental advocate Barry Commoner, in his 1971 book *The Closing Circle,* noted that between 1940 and 1970 the amount of phosphates in city wastewater increased from 20,000 to 150,000 tons per year. Phosphorus is a critical factor for plant growth in the water as well as on land. Wastewater, with its phosphorus-rich load of urine and feces, had long fertilized the receiving waters, and when the phosphates in detergents were added as well, the algae began to grow and grow and grow. Organic wastes provided so many nutrients—so much food—that the aquatic ecosystem became overwhelmed. On the bottom layers, the water was quickly depleted of oxygen, while anaerobic bacterial growth made the bottom muck stink and bubble. On the upper layers, the algae grew so splendidly on excess phosphorus that little else survived.

Continentwide, industries and cities were not the only polluters of the waterways. Agriculture adopted chemical fertilizers, pesticides, and herbicides; coal mines drained sulfuric acid into creeks and streams; and cyanide was used as a reagent to leach metals from low-grade ore.

As the demand for coal increased in the first decades of the twentieth century, the number and size of the coal mines grew. When coal surfaces are exposed, the sulfur in the coal comes in contact with water and air and forms sulfuric acid. As water drains from the mine, the acid moves into the waterways, and as long as rain falls on the mine tailings the sulfuric acid production continues, whether or not the mine is still operating. If the coal is strip-mined, the entire exposed seam leaches sulfuric acid. Strip mining was economical, but it rearranged the watershed, leaving the infertile subsoil on the surface and polluting streams. In 1939, West Virginia became the first state to require the reclamation of strip-mined land, and others followed.

By the late 1930s, it was estimated that American coal mines were producing about 2.5 million tons of sulfuric acid annually. Mine drainage was a particular problem in the Ohio River Basin, where twelve hundred operating coal mines drained an estimated annual 1.5 million tons of sulfuric acid into the waterways in the 1960s, and thousands of abandoned coal mines leached acid as well. The waterways were acidified wherever coal had been mined, though, and in Pennsylvania alone, mine drainage had blighted 2,000 stream miles by 1967.

Farming, like mining, replaced labor with machinery and fossil fuels, and then added a chemical arsenal: fertilizer displaced land, herbicides displaced the cultivator, and insects were controlled by synthetic insecticides. Farmers had always added silt to the waterways, and in the postwar era they added chemicals as well.

Before the 1940s, farmers had used their wits to maintain soil fertility and contend with disease and pest cycles. (Prayers and rituals also played a part in good harvests. As recently as the 1900s, some Plains farmers sowed their wheat fields naked, under a full moon. The farmer's wife would stride through the furrows in front of her husband, who threw handfuls of seed on her rump by the light of the moon.) They rotated their fields between cool-season plants, like wheat, rye, flax, and oats, and warm-season plants, like buckwheat and millet; between deep-rooted and shallow-rooted plants; between broadleaf and grassy plants; between cash crops of grain and soil-building legumes, like hairy vetch.

Pesticides released farmers from their ancient battle against insects. Chemical fertilizers provided a cheaper and more concentrated source of nutrients than manure, and crop rotation was no longer necessary. Herbicides took care of the weeds that competed with crops for nutrients and light. With new high-yield varieties, chemical helpers, irrigated fields, and oil-fueled machinery, big harvests were nearly guaranteed. Corn, a heavy user of nitrogen, could be grown in a field year after year. For a short time, it seemed as though food production no longer depended on nature at all.

Some of the drawbacks of industrialized agriculture became apparent quickly; others took decades to appear. The downside of pesticides showed up first. There was DDT's unnerving tendency to bioaccumulate, and then there was the high adaptability of insects, which lay lots of eggs. A few insects would invariably survive a toxic wipeout and produce resistant offspring, which would quickly repopulate the area. Pesticides also killed the beneficial insects that fed on the pests, removing natural checks from the ecosystem. And the poisons were washing into the waterways. In the 1960s, pesticides were reformulated to degrade more quickly, and pesticide use in the United States is currently between 250,000 and 375,000 tons a year—about 2 or 3 pounds per person—while the total crop losses to insects are approximately what they were before pesticides began to be used.

Chemical fertilizers provided no organic inputs to the soil, which began to lose tilth—that is, the soil nutrients, texture, and ability to hold moisture declined. Without the addition of crop residues and animal manures, the organic matter in the soil was consumed and not replaced. As the soil compacted, chemical fertilizers had to be used in higher doses. Illinois farmers used 11,000 tons of nitrogen fertilizer per USDA unit of crop production in 1949; two decades later, they were using 57,000 tons to produce the same yield. The efficiency of nitrogen combining with the crop declined fivefold, and fertilizer nitrogen that did not enter the crops ended up in the ecosystem. In the West in particular, where water was impounded and diverted to grow food in former deserts, water used for irrigation would return to the rivers and streams laden with salts, minerals, and agricultural chemicals.

By the 1960s, the United States waterways had been thoroughly disrupted by dams, channels, diversions, and dredging. The watersheds were missing the wetlands and forests that once had cleaned the water; the riparian zone was often bare; and rangelands that had bled water into the groundwater now hemorrhaged soil into the streams. The aquatic ecoystems were stressed in every direction, the water cycle was unable to cope with the onslaught of chemicals and organic matter, and rivers and lakes began to die.

Lake Erie, the shallowest of the Great Lakes—which were called the Sweet Seas by the early French explorers—was especially impacted by chemical and organic wastes. Along the north coast, silt and agricultural chemicals were swept into the water. Along the south coast, Buffalo, Cleveland, Toledo, and Detroit were dumping hundreds of millions of gallons of raw sewage a day, along with equally impressive amounts of industrial waste.

By the mid–1960s, every major river flowing into the lake was grossly polluted. The Buffalo River was laced with oil, phenols, color, organic matter, iron, acid, sewage, and various exotic organic compounds. Coliform counts reached 500,000 per 100 milliliters of water, and even sludge worms and leeches could not live in the industrially contaminated bottom sediments. The Detroit area added 20 million pounds of contaminants to Lake Erie every day, in a wastewater flow of 1.6 billion gallons. The Cuyahoga, which flows to the lake through Akron and Cleveland, ran chocolate brown on some days and rust red on others; it was so clogged with floating logs, oil, old tires, paints, and chemicals that the city of Cleveland declared it a fire hazard, and in 1969 it actually ignited, burning two railroad bridges in a toxic inferno of flames some 200 feet high.

All the Great Lakes were polluted, but Lake Erie, with the least volume, was the worst off. The first inkling that the Great Lakes were too small to absorb unlimited quantities of chemical and organic waste came in 1953. The limnologist Wilson Britt, while taking routine bottom samples in the west end of Lake Erie, found that the mayfly larvae were gone. Where they had formerly numbered between four hundred and a thousand per square meter, the populations now ranged from forty or fifty per square meter to zero. In their place were sludge worms, primitive segmented annelids called oligochaetes, which are creatures adapted to an oxygen-poor environment. Five years later, a 2,600-square-mile area of the central lake was found to be completely anoxic: there was no oxygen in the bottom waters at all. Algae formed mats up to 2 feet thick, covering hundreds of square miles of the lake's surface. Carpets of thousands of acres of algae would break off and wash up onshore, fouling water intakes and beaches.

The effects of Lake Erie's pollutants were soon spreading throughout the region. The insecticide Mirex, a chemical relative of DDT which was used to combat the spread of fire ants in the Southwestern United States, had been manufactured near Buffalo in the 1950s, and some had spilled into Lake Erie. It seemed secure in the bottom sediments until the early 1960s, when beluga whales in the St. Lawrence estuary started washing up dead—of Mirex poisoning, mostly, though their organs were said to be so loaded with toxics that the belugas were technically hazardous waste. As it turned out, when the female eels of Lake Erie returned to the Sargasso Sea to breed, they swam there via Niagara and the St. Lawrence estuary. The eels' migratory reserves of fat were laced with Mirex and other chemicals. The beluga whales in the St. Lawrence estuary ate an average of 200 pounds of eels apiece per year, and many of the whales died. Mirex that was spilled into the waterways years ago is still cycling in the environment.

By the 1960s, the watersheds surrounding Lake Erie had been debeavered, deforested, and plowed, the waters had been defished, the rivers to it had been dredged, straightened, and dammed. The flows of municipal and industrial wastes were unabated, and this Sweet Sea was sweet no longer.

As it was in Lake Erie, so it was in waterways across the country. Nonetheless, as the fifties closed, the degradation of the natural environment was of little concern to the United States government. The business of America was business, and business was good. Late in his second term, President Dwight Eisenhower established a commission to set goals for America in the upcoming decade. The commission's report, delivered in November 1960, was 372 pages long, but water pollution and air pollution were covered in just five paragraphs. There were fifteen goals for national concern and action in the 1960s; the environment was not one of them.

And yet, with the growth of the suburbs, the public anxiety over nuclear testing, and the subsequent public education on chemical transport in the environment, the groundwork for the environmental movement had already been laid.

In the postwar prosperity of the 1950s, blue-collar workers moved to the suburbs and bought cars. The automobile made wilderness activities like hiking, fishing, and nature photography more accessible, while pleasure boats and summer camps were no longer the province of the wealthy. As the population at large was rediscovering the delights of the natural world, the superpowers were testing their bombs. By 1951, the United States had detonated sixteen nuclear bombs and the Soviets thirteen; the following year, the British joined in with their first nuclear test. The explosions took place in remote locations, the results were shrouded in secrecy, and the newly formed Atomic Energy Commission would issue terse reports advising the citizenry that a test had occurred, and that the consequent radioactivity had been confined and was in any case harmless to the public.

In 1953, a cloudburst in Troy, New York, thousands of miles from our bomb test sites, was discovered to be highly radioactive. The discovery was an accident: physicists in a nearby university laboratory who were conducting experiments with radioactive material noticed a sudden surge in their background counts. There was no public report—that would have violated government secrecy rules—but physicists across the country started testing privately for radioactivity. It was everywhere. Rain, soil, food, and water had been contaminated by windblown fallout from nuclear test explosions. Radioactive fallout made its political debut in the 1956 presidential campaign between Adlai Stevenson and Dwight Eisenhower, and it was a central issue. Scientists and civic leaders began speaking at PTA meetings, in church, and to civic groups, explaining how radioactivity works. For the first time, the public at large began to understand that dilution was not necessarily the fate of pollutants, which can cycle through the biosphere in unexpected ways.

Fear of cancer, with its seemingly random onset, terminal effects, and lack of effective treatment, was another driving force behind environmental regulation. By the early 1960s, tox-

icologists were able to predict the effects of industrial and agricultural toxins on the general population, and epidemiological
studies confirmed that environmental pollution was linked
with the increasing incidence of cancer. No one knew (or
knows now) what fraction of cancers are caused by exposure to
toxins released to the environment, but water pollution began
to take on new significance. Untreated industrial pollutants in
the waterways were clearly more than an aesthetic insult: they
were a threat to life itself.

One of the environmental movement's prophets was Rachel
Carson, whose 1962 book *Silent Spring* described the effects of
DDT on bird's eggs in lyrical prose, and expanded on the issue
of chemical transport in the environment. If we continued to
dump chemicals into the environment, Carson argued, we were
likely to wake to a silent world. This book articulated the
movement's central doctrine: the earth faced an ultimate "eco-
catastrophe" unless immediate efforts were made to halt the
deterioration of the environment.

When the environmental movement began, it appealed prin-
cipally to well-educated, well-heeled white people. Environmental concerns correlated with nonrural residence, were
strongest among those fifty-five and older, and arose independent of political affiliation. At the time, it was said that environmentalists were people who had already bought their second
homes: these people had something worth protecting, and they
were old enough to remember a time when the waters weren't
polluted and the air was clean. They also had access to power
and understood the workings of the political process. Pollution
must have been cocktail conversation at the lawmakers' social
gatherings, for the first substantial water quality legislation—the
Water Quality Act—passed in 1965, requiring that water quality standards be established for all of the country's waterways.

The environmental movement was only beginning, but with
a skilled and literate core the cause spread quickly. Pollution,
which had formerly been seen as an inevitable by-product of an

industrialized society, was redefined as a social problem: The more goods we produced, the more we destabilized the environment. In 1969, the same year the Cuyahoga caught fire, the National Environmental Policy Act became law, intending to "create and maintain conditions under which man and nature can exist in productive harmony." The first Earth Day, April 22, 1970, was a nationwide event, with the aim of educating the public about the environment. Organizations such as the Sierra Club, the National Resource Defense Council, Greenpeace, the Nature Conservancy, the National Wildlife Foundation, and the Audubon Society had huge gains in membership. The environmental movement began to capture the imagination of a nation, mounted successful political campaigns, and eventually changed the daily habits of millions of individuals.

In 1972, the Federal Water Pollution Control Act—commonly called the Clean Water Act—was passed, which ruled that water quality was to be improved without regard to cost and set as a "national goal" the ending of the discharge of pollutants into the waterways by 1985. Every city in the country was required to build and operate a wastewater treatment plant, and the newly formed Environmental Protection Agency (EPA) provided grants and technical assistance. The waterways were far from the only focus, though. Other environmental laws enacted included the Clean Air Act; the Resource Conservation and Recovery Act (aimed at hazardous waste); the Toxic Substances Control Act; the Federal Insecticide, Fungicide, and Rodenticide Act; the Safe Drinking Water Act; and the Comprehensive Environmental Response, Compensation, and Liability Act, which worked to clean up old hazardous waste sites. It was now the law of the land that environmental, aesthetic, and social impacts be considered along with economic impacts for almost any type of proposed land, water, or resource development. Rarely in history had the cry of the wilderness been heard so keenly, felt so deeply, and acted on so decisively by so many. The waterways, clearly, were saved.

ELEVEN

WHAT SLUDGE
TELLS YOU

The Clean Water Act set curbs on discharges of all types, but since most of the problem was perceived to have been caused by cities and industries, most of the money was focused on reducing pollutants from these sources. In the decade after passage of the Clean Water Act, the federal government disbursed billions of dollars in wastewater-treatment-plant construction grants, and industries spent countless more billions on in-house solutions to clean up their effluent. Industry by industry and city by city, discharge reduction and source reduction were implemented, treatment plants were built, and toxic flows were reduced.

There were noticeable improvements. Fish kills declined, algal blooms began to disappear, and rivers and lakes around the country began to revive. The banning of DDT allowed the populations of ospreys, bald eagles, and brown pelicans to rebound, and recycling matured from a fringe activity to standard behavior. One of the most obvious effects of the new environmental laws was that the coal-fired plants that once belched great gobs of black smoke were required to clean up their smokestack emissions. When the particulates were removed, the results were completely unexpected.

The larger particulates in smoke—mostly ash—are chemically basic. When the smokestacks were low and black smoke blanketed a city, the basic particulates neutralized the acids that form in the air from the nitrogen oxides and sulfur oxides released during combustion. The rain was dirty, but the pH was probably neutral; even London, known for its black shroud, had healthy trees and gardens. In the 1940s, taller smokestacks were built to raise the soot above the cities, and this is thought to have been the start of acid rain. The particulates still fell on the surrounding communities, but the acids circulated high and were swept northeast by the prevailing winds. When the Clean Air Act passed and the smokestacks of the Rust Belt were no longer allowed to blast wastes into the air, the largest particulates were removed by filters, while the sulfur oxides and nitrogen oxides went straight out the chimney. In the Northeast, the rain was soon as acidic as orange juice.

As long as the pH of soil is neutral or basic, the minerals and trace metals that lace the soil are bound in place. The soils of the Northeast were barely neutral to start with, and after three decades of increasingly acid rain, the soil was unable to buffer the onslaught. The metals that had been bound in the soil moved into solution, and fish began to die from metal poisoning. Amphibians, with their permeable skin, were directly affected by the acids, as were high-altitude trees, which were bathed in acid fog. By the early 1980s, trees on the summit of mountains in upper New York and Vermont had begun to die, and the mountain ponds in the region were often fringed with blanched corpses of salamanders and frogs. More equipment on every smokestack followed, but abatement technology was not the whole story. Energy conservation, improved combustion efficiency, changes in products and in the means of production soon contributed to cleaner skies and sweeter rain.

In 1987, fresh out of MIT with a master's degree in Technology and Policy in hand, I was persuaded that we could engineer our way to clean water as well. And so I went to work on the last of the big-city wastewater projects: the Boston Harbor Clean-Up.

Boston has one of the oldest sewer systems in the country, and one of the most sprawling. In the 1860s, the city built holding tanks on Moon Island in Boston Harbor, and released the sewage on the outgoing tides. At the time, this was a thoughtful step forward in sewage management, and Boston's innovative system was written up in civil engineering journals here and abroad. Over a century later, however, little had changed but the number of towns connected to the wastewater collection system. Two primary treatment plants, sited on small former islands south and north of the city, received the wastewater from forty-three communities, including over six thousand industrial dischargers and 2.3 million people. And the sludge was still being released on the outgoing tide every day. The brown slick was supposed to drift out to sea, but more often it ended up circling back to shore instead. And along with the sludge, the prevailing tides washed ashore thousands of pink plastic tampon applicators.

In 1982, the Quincy city solicitor was jogging along a beach near Boston and realized that he was running in turds. Within a year, the city of Quincy sued the Boston municipal commission that managed the wastewater treatment plants. Soon the EPA joined in the fray with another lawsuit, and it was ruled that a secondary wastewater treatment plant would be built and that sludge would no longer be dumped into Boston Harbor.

I worked for the Massachusetts Water Resources Authority, which assigned five of us to divert the treated wastes to some sort of application on land. The disposition of the sludge awaited the siting and construction of a sludge-pelletizing plant, but the court had ordered that the scum be taken care of immediately. Before we get down to the details, though, we need to go deep into the innards of a wastewater treatment plant.

The tour starts in the lobby. Municipal funds for dressing up such unglamorous budget items are scarce, so wastewater treatment plants across the country are remarkably similar: the cheap linoleum and steel desks are as ubiquitous as the aquarium in the entryway and the whiff of digesting sludge from the works outside. The main that brings the raw sewage into a wastewater treatment plant is often enormous. In the United States, per-capita wastewater production is about 100 gallons a day, so city flows become epic. In Boston, it's greater than the combined flow of the three rivers—the Charles, the Mystic, and the Neponset—that feed into the harbor.

Wastewater is nearly all water, containing just 1 or 2 percent solids, including excrement and food, toilet paper and detergents, soaps, shampoos, cleansers, household hazardous waste, and oil and grease. The water is gray, and festooned with shards of toilet paper. A series of preliminary screens separate the socks, rags, sticks, and lumber from the sewage and protect the plant's pumps from debris. Next, the wastewater passes through chambers designed to slow the flow enough for grit to settle to the bottom. The grit is mostly sand and gravel, but also includes melon seeds, coffee grounds, cigarette butts (and, within a very short time, wiggly white worms); both the grit and the screened debris are usually turned into landfill.

The next stage of cleaning is done in giant rectangular open-air primary sedimentation tanks, or clarifiers. A blade on rails moves slowly across the surface of each tank, skimming off scum and shunting it to thickening tanks. Seagulls whirl above, and dive down to fish out vegetables and tampon applicators, which litter the walkways between the tanks. Clarifiers hold the wastewater for an hour or two, which is long enough for the flow to separate into layers. The scum floating on top of the tanks includes grease, soap, skin, vegetable and mineral oils, some paper, wood, and cotton, along with most of the Band-Aids, condoms, and plastic tampon applicators people flush down the toilet. (These applicators are a wastewater engineer's nightmare. Aerodynamically designed, they slip through screens like a torpedo. In the 1980s, roughly fifty thousand applicators a day were arriving at the wastewater treatment plants in Boston. This translates to a flush rate for these items of about 40 percent, though Playtex, their principal manufacturer, claimed at the time that informal surveys indicated only 1 percent of their customers flushed the applicators.)

Meanwhile, the sludge—mostly feces, toilet paper, and the heavier suspended solids—sinks to the bottom of the clarifier tank, where it is scraped into a sump or a hopper and sucked out the bottom of the tank for further processing.

After primary treatment, the water has dropped about two-thirds of the solids it carried, but it still contains some suspended, colloidal, and dissolved solids—urine, for example. Before Boston built its secondary treatment plant, this water was chlorinated and released to the harbor. During secondary treatment, the dissolved solids in the water are eaten by a complex community of microorganisms that are already part of the flow. The temperature, oxygen level, and contact time are controlled to maximize the growth of a slimy, bulbous bacterial sludge, which looks like it would be ready to walk out of the refrigerator in different circumstances. The effluent itself is chlorinated and discharged to the waterways. Secondary treatment routinely removes about nine-tenths of the solids in wastewater, and 95-percent removal is not unheard of. Tertiary treatment, including filtration or chemical precipitation (in which chemicals are used to flocculate the dissolved solids), can remove over 99 percent of the solids in water. This effluent is drinkable (if you are so inclined), but tertiary treatment costs are generally considered much too high for such a marginal improvement, and only a handful of cities today have such systems in place. Ninety percent cleaner is pretty good, but secondary treatment is no cure-all. One-tenth of anything you put down the drain or the toilet or into the washing machine sneaks through the treatment plant and out to the waterways within hours.

Meanwhile, we've left the residuals still in progress back at the treatment plant. The scum is concentrated in the thickening tanks, where the water is drained off and piped to the head of the plant, leaving mostly grease behind. The sludge is thickened as well, and then these residuals are both piped to an enormous covered vat called a digester, in which the raw sludge brew—a pungent, dark brown liquid the consistency of pancake batter—ages and ripens. It is kept warm, to encourage the growth of microorganisms, which consume—digest—the solids and give off methane. In most wastewater treatment plants (Boston's included), this methane is collected and used to provide part of the plant's power. A digester usually holds the sludge for at least three weeks, so large treatment plants need several large digesters.

In a typical two-stage anaerobic digester, a half million gallons of sludge is kept well mixed during the first stage, maximizing the contact of the organic matter with the microorganisms to encourage their growth. In the second stage, the mixer is turned off, and the sludge stratifies into four layers. Scum and odd chemicals rise to the top, and the comparatively clear supernatant below is piped back to the head of the plant for reprocessing. The actively digesting sludge burbles below all this, and the digested sludge—with most of the organic matter consumed by bacteria—sinks to the bottom, where it is pumped out of the digester to a holding tank.

The grease in the scum is digested more slowly than the organic matter in the sludge, so a scum blanket breaks down steadily as it accumulates. The plastics in scum, though, are less tractable. Bacterial digestion usually reduces condoms to an innocuous-looking ring easily mistaken for a rubber band, but the plastic tampon applicators have great structural integrity, and in a few years a layer of them 6 to 8 feet thick builds up, and the digester has to be emptied and cleaned.

When bacteria and other microorganisms have consumed most of the organic matter in the water and the sludge is well digested, the result resembles composted cow manure mixed with water. The phosphorus in digested sludge once polluted the waterways, but it is a valuable fertilizer on land, while the organic material and the microorganisms build up the soil biota. Digested sewage sludge was routinely reused as fertilizer until the 1940s, when cheap chemical fertilizers became available. For the next forty years, sewage sludge was used as landfill, or else it was incinerated or dumped back into the waterways. But after the Clean Water Act curbed industrial contributions to sewers, sludge began to be used on land again. Seattle sprayed its sludge on forests to promote tree growth, while composted sludge from Maryland was used to fertilize the White House lawn. New York City continued to fill up barges and dump its sludge at sea, however, and Boston continued to release its sludge on the outgoing tides. Not until 1988 did Congress prohibit the release of sludge to the ocean, and the ocean dumping of sewage sludge in the United States finally ended in June 1992.

For two years—from 1988 to 1990—Boston's scum was mine, but there was nothing to do with it. Landfill equipment slipped on the grease, so landfills would not accept it; incineration works poorly with a feedstock of grease and water; the fats were laced with sewage and useless for animal feed or soaps. Disposing of scum as a liquid was hopeless, so we made it into a solid. Every day, thousands of gallons of thickened scum were mixed with cement kiln dust, another waste product. The chemically fixed scum was a gray material encrusted with personal hygiene items. It behaved like a soil, mechanically, and eventually a few hundred thousand cubic yards of fixed scum was stored in three bulldozed hills. The rats and gulls would congregate there to eat; perhaps they liked the fats. My scum piles were ultimately incorporated into one of the berms designed to hide the new secondary treatment plant from view—a fitting use, I thought, for the concrete evidence that people won't change their behavior until they understand that their toilets are connected to the waterways.

Along with managing the scum, my job was to assess sludge quality to make sure that we met the EPA regulations for any sludge products that would be used on land. Nearly all the toxins and heavy metals in wastewater are transferred to sludge and scum in the wastewater treatment process. The chemical contaminants are not destroyed; some volatilize into the air, but most chemicals and heavy metals bind to either organic matter or grease, and end up in the digested sludge. Secondary treatment transfers 92 percent of PCBs in wastewater to the sludge, for example, along with 90 percent of the dimethyl phthalate and 70 percent of the cadmium. This means that digested sludge carries the fingerprint of all of the industrial discharges to the wastewater collection system.

Every city of fifty thousand people or more is therefore required by law to establish discharge limits for its industrial dischargers, a requirement that was initially ignored. When sludge started to be reused, though, municipalities got more involved in what arrived at the treatment plant. As soon as a city starts recycling its sludge, the industrial-discharge permit system tightens up. The inspectors pay closer attention, and an industry's furtive slips down the drain tend to decrease. A forty-city sludge survey in 1982 and the National Sewage Sludge Survey of 1988 both showed that cities that recycled their sludge had cleaner sludge, regardless of the number of industrial dischargers. According to the 1988 sludge survey, 13,458 waste-water treatment plants were generating nearly 6 million dry tons of sludge a year—or roughly 50 dry pounds of sludge per person annually. Of that, over a third was land-applied, an amount that was steadily increasing.

This tale of the dwindling industrial contaminants in the wastewater grew even stranger, though, because not all the industrial contaminants in sludge get there because of industrial discharges. When rain falls on pavement, it sweeps oil, gasoline, and the detritus of tire treads down into the sewers. Households add detergents, soaps, cleansers, household hazardous waste, lawn-care products that would never be bought if people read the small print, and oil that weekend mechanics dump into the gutter. The water itself from the reservoir has its own baggage of metal and other contaminants.

With municipal-sludge quality reports from around the country piled on my desk, I began to realize that industries themselves are no longer directly dumping much down the sewers. Whether or not industrial discharge reports are reliable, sludge implacably reflects whatever contaminants are present in the waste stream; sludge doesn't lie. And most city sludge is remarkably clean. The waterways, however, are still polluted.

What was dawning on me in the late 1980s because of local experience was beginning to show up nationwide. Water quality had markedly and rapidly improved in the 1970s, but by the mid-1980s a third of the nation's waterways were still assessed as unfit for fishing or swimming—and so they have remained. About 40 percent of the lake acreage and 30 percent of the stream miles in this country are still polluted. Water quality has changed little in the last decade.

Over half the rain that falls on the continental United States—roughly 35 billion gallons of water a day—is made into wastewater by our cities and our industries, and our cities clean it reasonably well. The overwhelming majority of water pollutants are now contributed by big agriculture. In 1992, the Executive Office's Council on Environmental Quality attributed 6 percent of the impairment in streams and lakes to industrial pollution and a whopping 60 percent to silt and excess nutrients from fertilizer runoff. In 1992, the U.S. Public Interest Research Group, a nonprofit environmental organization, estimated that toxic industrial contributions to the waterways and oceans totaled about 155,000 tons—or less than 5 percent of what Detroit alone used to contribute every year in the 1960s. In the meantime, however, tens of millions of tons of topsoil wash into the streams, and farmers dump about 20 million tons of fertilizer on fields across the country, along with between 250,000 and 375,000 tons of pesticides. Agricultural subsidies—which were designed in the 1930s, before the advent of chemical fertilizers and pesticides—are at the heart of the problem. These subsidies, which are now being phased out, allow farmers to grow more crops than would be grown in a consumer-driven market, and in order to maximize their yields many farmers turned to the use of fertilizers and pesticides. Fortunately, organic farming techniques and biological (rather than chemical) pest management are gaining ground. Over a third of the nation's corn crop is now grown using low-till or no-till farming, in which the crop waste is recycled into the soil and earthworms create macropores, which reduce runoff and increase water uptake by the soil. The high costs of chemical fertilizers and pesticides—and the premium that consumers are willing to pay for organically grown crops—is changing the way many farmers manage their land.

But some types of pollution are far more difficult to manage. Groundwater pollution has proved to be surprisingly intractable; once a plume of contaminants is loosed into an aquifer, it is nearly impossible to remove. Likewise, airborne pollutants continue to rain down steadily from the skies even though the air has grown clearer. Remote ponds that appear to be pristine are polluted by assorted wind-deposited chemicals and metals, including mercury from incinerator smokestacks, DDT from Central American fields, and traces of other assorted hazardous wastes. Finally, we cannot undo what we have done. New hazardous waste sites are not being created, but the old ones are very costly to clean up; PCBs, once they have been released to the environment, cannot be taken back.

Cleaning up industrial and municipal discharges has cost hundreds of billions of dollars, and a third of the nations's waterways are still polluted. Water pollution is clearly more complicated than we had realized, and discharge control has not solved the problem.

Consider this: after water leaves a reservoir, it takes hours or days to move through the pipes to a house or factory, where it's polluted; through a wastewater treatment plant, where it drops most of its pollutants; and out to the waterways or to the sea. After days running through the engineered system, it enters the natural water cycle for a decade or more, where it may run to the sea and eventually move to the clouds, be blown inland to rain onto forests and fields, run into wetlands and streams, or percolate down to the groundwater. And then, perhaps, it moves into a well or reservoir again. For each day that water flows in pipes, it might spend a decade or more in the natural world. And nature is the best cleanser—at least, it once was.

This country's waterways have been transformed *by omission*. Without beavers, water makes its way too quickly to the sea; without prairie dogs, water runs over the surface instead of sinking into the aquifer; without bison, there are no ground-water-recharge ponds in the grasslands and the riparian zone is trampled; without alligators, the edge between the water and land is simplified. Without forests, the water runs unfiltered to the waterways, and there is less deadwood in the channel, reducing stream productivity. Without floodplains and meanders, the water moves more swiftly, and silt carried in the water is more likely to be swept to sea.

The beaver, the prairie dog, the bison, and the alligator have been scarce for so long that we have forgotten how plentiful they once were. Beaver populations are controlled, because they flood fields and forests, while wetlands acreage decreases annually. Prairie dogs are poisoned, because they compete with cattle for grass, while the grasslands grow more barren year by year. Buffalo are generally seen as photogenic anachronisms, and alligators are too reptilian to be very welcome. But all of these animals once shaped the land in ways that improve water quality.

It is not only water quality that has been affected, though. Without these builders, the contour of the land has been smoothed, and many niches have disappeared; the base of the food chain has withered, and in the process the abundance and the productivity of the land has declined. In 1993, the National Biological Service was created by Secretary of the Interior Bruce Babbitt to gather biological data on public and private lands nationwide. Babbitt, a geologist by training and a former two-term governor of Arizona, was eager to bring ecosystem management to federally owned lands. In 1995, the National Biological Service completed the first ecological review of the United States. The study found that the extent and vitality of dozens of ecosystems throughout the country have suffered a sweeping, but largely unnoticed decline. In spite of gaps and uncertainties in the data, the researchers concluded that the information "portrays a striking picture of endangerment."

The tallgrass prairies, the oak savannas bordering the grasslands, and the old-growth, fire-managed deciduous forests of the Eastern United States are among the largest imperiled ecosystems, along with more than 100,000 square miles of long-leaf pine forests that once covered the Southeastern coastal plain. The Eastern forests were cut, the Midwestern grasslands have nearly disappeared under crops, and the oak savannas have been degraded by fire suppression. The longleaf pine ecosystem has faced both types of decline: great swaths were cut down in the early part of the twentieth century, and without periodic fires the pines have been taken over by hardwoods.

Cattle and sheep have destroyed much of the riparian edge in the grasslands, degrading the aquatic ecosystems. When a stream has a well-developed edge, it supports life that cleans the water. In the mixed-grass and short-grass prairies, cattle trample the edge of the stream. Without beavers to cut down or drown the older streamside trees, the cottonwoods become enormous, siphoning water from the flow. What used to be a live stream, with fish and a lush edge, becomes a barren gulch. The water table drops, erosion progresses, and in fifty or a hundred years the land becomes desert. In 1988, an Arizona Fish and Game Department report concluded that less than 3 percent of the state's original riparian zone remained intact, while New Mexico has lost at least 90 percent of its riparian zone to grazing.

According to the National Biological Service's 1995 report, ecosystems that once covered at least half the area of the contiguous United States are now critically endangered. The face of the land has changed, and as ecosystems are fragmented and simplified, the species that rely on them for habitat are under increasing assault. The Nature Conservancy's 1996 Annual Report Card for United States Plant and Animal Species, a comprehensive assessment of the country's indigenous fauna and flora, has found that the mammals and birds are doing relatively well; however, flowering plants and freshwater species (like most mussels and many riverine fish) are not. Of 20,481 native species of plants and animals surveyed, about one-third are faring poorly. Over two hundred and fifty species are extinct or possibly extinct, over thirteen hundred are critically imperiled, eighteen hundred are imperiled, and over three thousand are considered vulnerable.

In the last century, twenty-one species of freshwater mussels and forty species of freshwater fish became extinct. Today, about two-thirds of the remaining freshwater mussel species in this country are at risk of extinction, along with about one-third of the species of amphibians and freshwater fish. All depend on rivers, streams, or lakes, which are becoming biologically poorer even as water quality improves. In their Annual Report Card, the Nature Conservancy attributes part of this to the long-term effects of dams and other water diversions.

When the Federal Energy Regulatory Commission (FERC) first issued fifty-year operating licenses for dams in the 1940s and 1950s, rivers were managed to provide energy production, irrigated acreage, and city water supplies. Life in the river itself and along its banks and on its floodplains was not included in the equation. Dams cause periodic and drastic fluctuations in the channel flow, as water is released to generate electricity for peak power demands. Fish are cut off from their spawning grounds or killed in the electric turbines. Dams block the flow of nutrients, slow and heat the water, and use the kinetic energy that would reaerate the water to generate electricity. In the last twenty years, however, our view of a river has expanded to include recreation and aquatic habitat, and dams are beginning to be managed differently. In 1986, the FERC was required to give recreational and biological issues equal consideration with power generation whenever it renewed a dam's license or issued a new one. In many cases, so little of a dammed river's water had been dedicated to the aquatic habitat that a small decrease in power production resulted in huge gains for wildlife. On the Deerfield River in Massachusetts, for example, a 1996 agreement to decrease power generation by 10 percent is expected to result in a fifteenfold increase in the trout habitat.

The movement to give rivers back their form was spearheaded by whitewater enthusiasts, but soon reached many environmental and governmental organizations. In 1991, the Department of the Interior ordered that dams in Wyoming, Utah, and Colorado release water flows on the Colorado River in the spring and early summer to enhance river rafting, rather than using the water to generate power at peak usage times. On March 26, 1996, a week-long artificial flood was released from the Glen Canyon Dam, and the Grand Canyon was scoured by high spring flows after thirty-three years of low water. As a result, for the first time since it was dammed, the Colorado River is beginning to regain its natural flow. About eight hundred dams nationwide are up for relicensing between 1996 and 2010, and the waterways should eventually show the difference.

When more water is allocated to maintaining a river's ecosystem, there is, obviously, less water available for cities, industry, and agriculture. Demand, though, has proved to be more flexible than people had imagined. Agriculture consumes the largest fraction of Western water by far—some 90 percent of it—and installing more efficient irrigation systems and growing crops that require less water can reduce water consumption considerably. Industrial and municipal water consumption are both rate-sensitive, and when the price of water climbs the purchase of low-flow appliances increases, another way of reducing demand.

The Army Corps of Engineers—past masters of dredging and channelization—have also become more environmentally aware. In 1969, the Corps was branded Public Enemy No. 1 by Supreme Court Justice William O. Douglas for its environmentally devastating activities. In southern Florida in particular, the Corps projects were so destructive that they are finally now being undone.

The Kissimmee River originally meandered south for 140 miles before it drained into Lake Okeechobee. South of the lake lies the Everglades, a great sheet of water supporting a river of grass that once extended unbroken 150 miles to the southern tip of Florida. In 1928, a flood in south Florida drowned 2,750 people, and the Corps avenged their death by trammeling the Everglades. The sinuous curves of the Kissimmee were straightened into a 56-mile canal 175 feet wide and 30 feet deep. An immense earthen levee was built around Lake Okeechobee, and 1,400 miles of canals, levees, spillways, and pumping stations were built to free up over half of the total Everglades marshlands for agriculture. Land speculators, cattle ranchers, sugarcane growers, and other agriculturalists grew rich and powerful, but by the 1960s the river and lake were polluted. The southern end of the Everglades (1.4 million acres) was left wild, but too little water was allocated to maintain its ecosystem. By the 1970s, the former hordes of waterfowl had dwindled to flocks, agricultural runoff had overfertilized the plants and poisoned the animals, and the Everglades began to dry up.

Today the Corps is working to free the Kissimmee from its dikes and locks, and to restore the wetlands that once cleaned the river. The original $372 million Corps of Engineers' project to restore 22 miles of canal into 43 miles of curvaceous Kissimmee has expanded into a $1.2 billion federal and state effort to restore the entire river. New water allocations to the Everglades are reviving the marsh, and people throughout the state are working together to ensure that Florida's paradise of grass is not lost. The Everglades is not an isolated case: in states across the country, the Corps now works in concert with the Audubon Society and other environmental organizations to manage water to maximize its benefit to wildlife.

After a hundred years of taking away the form of our waterways, we are starting to put some of the pieces back. This about-face will doubtless do much to help many aquatic species, but it may be too late to save most of the native freshwater mussels.

The zebra mussel (*Dreissena polymorpha*) is an ancient resident of the Caspian Sea that spread throughout the European waterways in the last two centuries. In 1985 or 1986, biologists believe that a ship from Europe jettisoned a number of larval zebra mussels along with its ballast water into the waters of Lake St. Clair, which is tucked between Lake Erie and Lake Huron. The larvae of the zebra mussels, unlike that of the fish-dependent American freshwater mussels, are free-floating. And while the American mussels burrow in the bottom sediments, the zebra mussels, like their saltwater counterparts, cling with tough byssal threads to such surfaces as water intake pipes and ship hulls. The native freshwater mussels have been in decline for a century, and the zebra mussels are spreading to fill the ecological void. This alien has reproduced so successfully that in some areas the sparse populations of native mussels are encrusted with pistachio-size, elegantly striped zebra mussels. Each small mussel filters about a quart of water a day as it gorges on microorganisms, and the water in the Great Lakes and the Mississippi River is finally beginning to be cleaned by filtering mollusks again. The native mussels, though, are unlikely to survive the competition.

The success of the zebra mussel could bring the Endangered Species Act to its knees. This continent is rich in species of freshwater fish and mollusks endemic only to a relatively few small and scattered places. The Mississippi River drainage system is a major center of endemism, as are the springs of the Southwest. As the populations of freshwater mussels, Texas blind salamanders, tiger salamanders, Lahontan cutthroat trout, humpback chubs, Colorado River squawfish, and Devil's Hole pupfish dwindle, one can't help but wonder; Do we have to save *all* of them?

It's the wrong question. By focusing on the preservation of endangered species one after another, as if they were items in a catalog, we are missing the larger, ecological picture. Without the ancestral complement of keystone species—nature's engineers—the path that water takes through the land, and the shape of the land itself, have been simplified. The central actors, which once numbered in the hundreds of millions and billions, are missing or scarce in most of their former range. Where do the salmon fingerlings hide and feed without the beaver ponds they coevolved with? How can the eagles prosper with no prairie dogs to eat? The tiger salamanders have no tunnels to hide in; the prairie chickens have no stage for their courtship dances. Without restoring the ancestral populations of engineers to at least some of the landscape, it seems unlikely that the supporting players will manage to survive.

Back when I was managing scum, I learned that before people are willing to change the ways in which they impact the waterways, they have to understand how the system works. People will continue to flush plastics down the toilet, for example, until they understand that their toilets are connected to the waterways. People will continue to keep prairie dog towns and beaver colonies off their land until they understand that the pathways water takes through the land are changed by tunnels and wetlands and that their land will hold more water, and grow more grass, with these animals. Ranchers in the Southwest are still poisoning phreatophytes that steal water from the river, and the beavers that once made wetlands in that arid region—and cut down the cottonwoods before they grew to be river-draining goliaths—are forgotten. It seems unlikely that enough private landowners will allow either rodent species back on their land to make a difference in the nation's water quality.

We have an enormous national commons, however. Over a fifth of the contiguous United States—626,000 square miles—is publicly owned and federally managed by the Bureau of Land Management, the Forest Service, the Fish and Wildlife Service, and the National Park Service. In theory, both the prairie dog and the beaver could be restored on all that vast acreage with a stroke of the pen. But while it is well understood that degraded land pollutes the waterways, public land is managed with little thought to water quality. For the last century, timber production, grazing, and oil and mineral extraction have been the primary uses of our national commons. Beaver populations are controlled to maximize timber revenue, while net wetlands loss continues; riparian habitat is stripped by livestock, lowering water tables and degrading water quality; careless lumbering clogs the streams with silt.

The U.S. Forest Service manages about 265,000 square miles of forestland in the contiguous United States (and a lot more in Alaska), or about 8 percent of the forty-eight states. Although forests are water purifiers, that capability has been largely ignored. In the West, vast clear-cuts allow silt to clog mountain streams, affecting salmon fisheries and municipal water supplies. Forests, instead of cleaning our water, are fouling it. By logging in ways that preserve stream integrity—for example, by avoiding the logging of steep mountainsides and by limiting clear-cutting—public forestlands could begin improving stream quality rather than impairing it.

The Bureau of Land Management, which manages our public grasslands, controls about 275,000 square miles of land (or about 9 percent) of the lower forty-eight. Some 23,000 livestock raisers lease this land and parts of Forest Service land as well, to graze a total of over 4 million cows and 2 million sheep. Some of the National Parks (which total about 75,000 square miles, or 2 percent, of the contiguous United States) support herds of buffalo, but the public grasslands of the Bureau of Land Management do not. There are 3,000 buffalo in Yellowstone National Park alone; in 1991, only an estimated 448 wild buffalo grazed on the bureau's 275,000 square miles. Aside from ripping up the water's edge, cattle and sheep are beset by predators in the forests and on the open range. Coyotes, bears, mountain lions, and wolves all prey on domesticated animals, and ravens pluck out the eyes of lambs. To keep the public lands safe for domesticated animals, the U.S. Department of Agriculture's Animal Damage Control Program destroys about 80,000 coyotes, 200 mountain lions, nearly 10,000 black bears, and 125,000 prairie dogs annually.

Without prairie dogs, beavers, or an intact riparian edge, the grasslands have barely survived. In 1991, the Executive Office's Council on Environmental Quality estimated that half the country's public rangelands were in poor or fair condition, while only 5 percent were judged to be in "excellent" condition—and because of recent efforts to reduce overgrazing and restore riparian habitat, this is the best condition the public rangelands have been for the past century. (Range in "fair" condition has between one-quarter and one-half the vegetation it should, while range in "excellent" condition has 75 percent or more.)

Grasslands need to be grazed, and cattle can be raised without degrading the land and water. Some Western ranchers are restoring riparian habitat by herding their cattle across the land to keep them from congregating streamside and by limiting foraging to specified parcels at particular times of the year. When sheepdogs protect a flock, predation drops sharply, and some ranchers are learning to live with more wildlife. Better still, buffalo are finally being raised for meat, and their populations are the highest they've been for a century. In 1993, a hundred and thirty thousand buffalo grazed, wallowed, and took good care of the water's edge; most of them lived on private ranchland, along with a hundred million cattle. In 1993, three private buffalo ranchers had larger herds than Yellowstone's. The country's largest buffalo herd is owned by media mogul Ted Turner, who recently also returned the prairie dog to his land.

Buffalo are excluded from public grasslands because they are carriers of brucellosis, a disease that has little effect on buffalo but causes cows to abort their fetuses. As the buffalo population has increased, though, it appears that brucellosis is far less contagious than had been assumed, and that buffalo pose little threat to the cattle with which they share the range. Buffalo meat tastes a lot like lean beef. And since buffalo evolved on the prairies, they are far hardier than cattle and can be raised without antibiotics, hormones, or artificial growth stimulants. In the winter, they sweep the snow away from the grass with their massive heads and shoulders and eat snow for water; they survive temperatures that freeze cattle solid. Ranchers who raise buffalo sell their meat to specialty restaurants and some of the better meat markets countrywide; the price ranges from about $6 a pound for ground meat to $20 a pound for steaks, and so far the demand has outstripped the supply.

The popularity of the prairie dog, however, has not improved. Now probably no more than 1 percent of the area once covered by prairie dog towns is tunneled. Although millions survive, their populations are vulnerable. The sylvatic plague is still endemic in prairie dogs, but the establishment of new colonies is restricted; without migration, *Pasteurella pestis* may get the best of the prairie dog. According to metapopulation theory, species are composed of networks of small, interacting populations that help to maintain one another: when populations decline, they can be rescued by healthy migrating neighbors. Healthy populations produce many migrants, thereby creating a positive-feedback loop resulting in a healthy population network. But if enough colonies falter, the network collapses. The sylvatic plague still depopulates prairie dog towns, and the migrants needed to repopulate them could become scarce.

Since the early 1980s, tensions have escalated between environmentalists and the spokesmen (mostly conservative politicians from Western states) for "endangered" loggers, mining companies, and ranchers. Large-scale logging, careless grazing, and corporate mining are still the main uses for our public land, but the environmentalists have succeeded in cordoning off nearly 52,000 square miles in eleven Western states—almost 5 percent of the West—as wilderness. The Northern Rockies Ecosystem Protection bill (not passed at this writing) seeks to set aside another 25,000 square miles in five areas, with wildlife corridors between them, which would put 1 percent of the country's old-growth forests under protection. Another 50,000 square miles of land—about the size of Iowa—has been taken out of agricultural production along the Mississippi flyway as part of the Conservation Reserve Program, which was set up in 1985. Ten years later, an estimated 83 million birds migrated south along the flyway, the biggest such migration in half a century.

There is the will to restore our land. We have just forgotten much of what is missing from it. The balance of nature that existed before we turned things upside down, and the richness and abundance of the land, were based on a few keystone species. What really matters are the numbers. In an area of 2.9 million square miles, billions of prairie dog tunnels and countless millions of beaver dams and buffalo wallows are significant. With their removal went an ecological system that cleaned the water and enriched the land. In spite of our earnest engineering efforts, about a third of the waterways are still polluted, and the natural water cycle is still hugely simplified. But some of the filtering mollusks and the buffalo are coming back, and the prairie dogs and the beavers have both survived with their culture intact.

This land once had clouds of birds, dense herds of grazers, myriad shoals of anadromous fish—and so it could again. It is time to restore the balance to our land and allow nature's engineers to do their work. If the prairie dogs and the beavers are allowed to reestablish their ancestral populations on public land, the dirt will fly, and the waterways will begin to regain their former pristine glory. On public land, at least, it is time for the beavers and the prairie dogs to come home.

NOTES

CHAPTER I: THE FUR TRADE

Page

3–4 The wearing of furs in the Middle Ages: Douglas Gorsline, *What People Wore: A Visual History of Dress from Ancient Times to Twentieth-Century America* (New York: Viking, 1952); R. Turner Wilcox, *The Mode in Furs: The History of Furred Costume of the World from Earliest Times to the Present* (New York: Scribner's, 1951); Robert Fossier, *Peasant Life in the Medieval West* (Oxford: Blackwell, 1988).

4 Skins bought by King Edward I: See Clive Ponting, *A Green History of the World*, 1991 (New York: St. Martin's Press: *The Environment and the Collapse of Great Civilizations*) p. 178.

4 For the price of new and used furs, see Elspeth Veale, *The English Fur Trade in the Later Middle Ages* (Oxford: Oxford University Press, 1966) p. 12.

4 Viking trade in furs: Ibid., p. 62.

4–5 Records of furs imported from Russia are in Robert Delort's *Le commerce des fourrures en Occident à la fin du Moyen Age* (Rome: École Français de Rome, 1978) p. 196; and in Veale, *English Fur Trade*, p. 69.

5 The reasons for the high demand for beavers: See Delort, *Le commerce des fourrures*, p. 181, and Joseph Reichholf, "Beavers," in Grizmek's Encyclopedia of Mammals, vol. 3 (New York: McGraw-Hill, 1990).

5 The statistics on waning of European beaver trade are from Delort, *Le commerce des fourrures*.

7 Decree of Charles I: See *Stuart Royal Proclamations, Vol. II: Royal Proclamations of King Charles I 1625–1646* (Oxford: Clarendon Press, 1983) pp. 613–618.

7 "The high-crowned Spanish beaver hat": Carolyn Merchant, *Ecological Revolutions* (Chapel Hill: University of North Carolina Press, 1989) p. 42.

7–8 In 1670, the Hudson's Bay Company was formed to trade with the Indians for fur, and the standard price for a beaver skin was set on the first expedition. Mari Sandoz, *The Beaver Men: Spearheads of an Empire* (New York: Hastings House, 1964) p. 91.

8 "The beaver does everything perfectly well": Quoted in A. Radclyffe Dugmore, *The Romance of the Beaver: Being the History of the Beaver in the Western Hemisphere* (Philadelphia: Lippincott, 1913) p. 188.

8 The history of syphilis is from Charles Panati's *Extraordinary Endings of Practically Everything and Everybody* (New York: Harper & Row, 1989) pp. 236–240.

9 Indian mortality rates from smallpox: William Cronon, *Changes in the Land: Indians, Colonists, and the Ecology of New England* (New York: Farrar, Straus & Giroux, 1983) p. 86.

11 The women "sold their coats from their backs": Quoted in William Bradford, *Of Plymouth Plantation 1620–1647*, ed. Samuel Eliot Morison (New York: Knopf, 1952) fn. pp. 89–90.

12 The number of ships trading off the Maine coast: Moloney, *The Fur Trade*, pp. 32–33.

12 "This year . . . the revenue is much diminished": Lewis H. Morgan, *The American Beaver: A Classic of Natural History and Ecology* (Philadelphia: Lippincott, 1868) p. 244.

12–13 The sustainable Canadian fur trade: David J. Wishart, *The Fur Trade of the American West, 1807–1840: A Geographical Synthesis* (Lincoln: University of Nebraska Press, 1979) p. 32.

13 "These animals are more prolific than our sheep": Quoted in Dugmore, *Romance*, p. 152.

15 List of tribes involved in trading: Sandoz, *Beaver Men*, pp. 147–148.

15 A single expedition: Ibid., p. 149.

15 "The beaver had been all but eliminated": Leonard L. Rue III, *The World of the Beaver* (Philadelphia: Lippincott, 1964).

15–16 Lewis and Clark's reports: Wishart, *Fur Trade*, pp. 18–19.

16 Two thousand Indian trappers: Ibid., pp. 190–193; and see also Alexander Ross, *Adventure of the First Settlers on the Oregon or Columbia River, 1810–1813* (London: Smith, Elder, 1849; Ann Arbor: University Microfilms, 1966) pp. 114–116.

17 This estimate of beaver population in the contiguous United States is based on telephone conversations with the furbearer wildlife biologist of

NOTES

each state. Since the beaver is not an endangered species, most of the state biologists could provide no more than an educated guess for their state's population.

CHAPTER 2: NATURE'S HYDROLOGISTS

Page

19–20 Beaver behavior as a pet: A. Radclyffe Dugmore, *The Romance of the Beaver: Being the History of the Beaver in the Western Hemisphere* (Philadelphia: Lippincott, 1913) pp. 172–173.

20 Indian legends of the beaver: See especially Mari Sandoz, *The Beaver Men: Spearheads of an Empire* (New York: Hastings House, 1964) p. 23; and Enos Mills, *In Beaver World* (Boston: Houghton Mifflin, 1913).

21 Estimates for the pre-Columbian beaver population range anywhere from 60 million to 400 million. I'm using the number 200 million as approximately mid-range. See also Ernest Thompson Seton, *Lives of Game Animals* (New York: Doubleday, Doran, 1929) pp. 447–448.

21–22 Castoreum and salicylic acid, and as a perfume base: Steffen Arctander, *Perfumes and Flavor Materials of Natural Origin* (Elizabeth, N.J.; published by the author, 1960) pp. 136–138. See also Jessica Maxwell, "Leave It to Beavers," *Audubon*, March–April 1994, pp. 104–109.

23 Working by night: The accounts of early North American trappers and explorers reveal that the North American beaver was diurnal (awake and working during the day) when the country was discovered and became nocturnal by the 1800s. Recent behavioral studies indicate that in some regions beavers are again becoming diurnal. My neighbor Jan Brodeur, in downtown Waitsfield, Vermont, watches beavers work during the day on their lodge in a wetland abutting her garden.

24 Beaver life: See Lewis H. Morgan, *The American Beaver: A Classic of Natural History and Ecology* (Philadelphia: Lippincott, 1868); Leonard L. Rue III, *The World of the Beaver* (Philadelphia: Lippincott, 1964); Morrell Allred, *Beaver Behavior: Architect of Fame and Bane* (Happy Camp, Calif.: Naturegraph, 1986); and Hope Ryden, *Lily Pond: Four Years with a Family of Beavers* (New York: William Morrow, 1989).

25 Aquatic fungi: Felix Barlocher, ed., *The Ecology of Aquatic Hyphomycetes* (New York: Springer-Verlag, 1992) pp. 1, 99.

25–27 Wetlands web of activity: Joseph S. Larson and Richard B. Newton, *The Value of Wetlands to Man and Wildlife* (Amherst: Cooperative Extension Service, University of Massachusetts, 1987); William J. Mitsch and James G. Gosselink, *Wetlands* (New York: Van Nostrand Reinhold, 1986); Brian Moss, *Ecology of Fresh Waters* (New York: Wiley, 1980) chapter 3.

30–31 Nineteenth-century trapping: James Bateman, *Animal Traps and Trapping* (Newton Abbott, Eng.: David & Charles, 1971).

NOTES

CHAPTER 3: THE WOODS

Page

35–36 Early settlers' attitudes toward forests: John G. Mitchell, "Whither the Yankee Forest?" *Audubon*, March 1981, pp. 76–99.

36 Josselyn quote: Charles F. Carroll, *The Timber Economy of Puritan New England* (Providence: Brown University Press, 1973) p. 59.

36 Marsh's voluminous footnotes are a glimpse into another era. On page 31, for example, he clarifies bog nomenclature by translating the eleven Lapp words for various types of swamps into Latin.

37–38 Firing of woodlands and grasslands by Indians: Stephen J. Pyne, *Fire in America: A Cultural History of Wildland and Rural Fire* (Princeton: Princeton University Press, 1982) p. 74. In addition to managing the forests, Indians would deform saplings to mark trails and springs, by bending the branches to point the way. An ancient oak on my great aunt Elise Kinkead's farm in Poughkeepsie was one of these Indian sentinel trees, with an enormous limb bent two hundred years ago to mark the path to a spring.

38 John Winthrop Jr. to Henry Oldenburg, Winthrop Papers, Massachusetts Hist. Soc. Coll., 5th ser. 8 (1882) pp. 124–125.

38 Increase in wildlife populations: William Cronon, *Changes in the Land: Indians, Colonists, and the Ecology of New England* (New York: Farrar, Straus & Giroux, 1983) p. 51.

40 Miantonomo's lament: "Leift Lion Gardiner: His Relation of the Pequot Warres," Massachusetts Hist. Soc. Coll., 1st ser. 3 (1833) pp. 154–155.

40 Colonial exploitation of timber: Carroll, *Timber Economy*, p. 25.

41–42 Timber use in the eighteenth and early nineteenth centuries: R.V. Reynolds and A. H. Pierson, "Fuel Wood Used in the United States, 1630–1930," *USDA Circ.* 641 (February 1942) pp. 9, 14.

42 Farmers cleared 31,250 square miles: John Perlin, *A Forest Journey: The Role of Wood in the Development of Civilization* (Cambridge: Harvard University Press, 1991) p. 355.

44 The microclimate of the redwood forests: Lorus Milne and Margery Milne, *Water and Life* (New York: Athenaeum, 1964) p. 112.

46 "Ink cannot tell the glow": Linnie Marsh Wolfe, *Son of the Wilderness: The Life of John Muir* (Madison: University of Wisconsin Press, 1945) p. 161.

47 Extent of the country's original forests: See Clive Ponting, *A Green History of the World* (New York: St. Martin's, 1991); and Gerald J. Grey and Anita Eng, "How Much Old Growth Is Left?" *American Forests*, September-October 1991, pp. 46–48.

49–50 Lichens in old-growth forests: Kevin Krajick, "The Secret Life of Backyard Trees," *Discover*, November 1995, pp. 93–101.

52 Dimensions of Vermont trees circa 1800: George Perkins Marsh, *Man and Nature, or Physical Geography as Modified by Human Action* (Cambridge: Harvard University Press/Belknap, 1965 [1864]) p. 236.

CHAPTER 4: THE VOYAGE OF RAINFALL

Page

56 Retention of organic debris: Chris Maser and James Sedell, *From the Forest to the Sea: The Ecology of Wood in Streams, Rivers, Estuaries and Oceans* (Delray Beach, Fla,: St. Lucie, 1994) p. 14.

57 Nutrients in the stream ecosystem: Geoffrey Petts, *Impounded River: Perspectives for Ecological Management* (New York: Wiley, 1984) p. 15.

57 Stream ecology: James V. Ward and Jack A. Stanford, eds., *The Ecology of Regulated Streams* (New York: Plenum, 1979); and National Research Council, *Restoration of Aquatic Ecosystems: Science, Technology and Public Policy* (Washington: National Academy Press, 1992).

59 Mississippi's deltaic lobes: *Restoration of Aquatic Ecosystems*, p. 400.

59 Flood renewal of fish hatcheries: Ibid., pp. 179–180.

60, 61 Army Corps of Engineers' records: See Maser and Sedell, *From the Forest to the Sea*, p. 123.

60 Islands of matted trees in the Indian Ocean: Ibid., pp. 105–106.

61 Species diversification on islands: See particularly Meredith Small, "The Seven Macaques of Sulawesi: Radiation on an Intermittent Archipelago," *Pacific Discovery*, Summer 1995, pp. 24–27; and David Day, *A Doomsday Book of Animals* (New York: Viking, 1981) pp. 27, 250.

62 The Darwin quote is found in Carolyn Merchant, *Ecological Revolutions: Nature, Gender, and Science in New England* (Chapel Hill: University of North Carolina Press, 1989) p. 159.

63 Water transpiration of a mature oak: Warren Viessman Jr., John W. Knapp, Gary L. Lewis, and Terence E. Harbaugh, *Introduction to Hydrology*, 2d ed. (New York: Harper & Row, 1977) p. 54.

64 Creation of springs: Richard M. Ketchum, *The Secret Life of the Forest* (New York: American Heritage, 1970) p. 15.

CHAPTER 5: A SEA OF GRASS

Page

67 The formation of grasslands: Carl O. Sauer, "Grassland Climax, Fire and Man," *Jour. Range Management* 3 (1950) pp. 16–21.

68 Shoot production by spring fires: David C. Glenn-Lewin, Louise A. Johnson, Thomas W. Jurik, Ann Akey, Mark Leoschke, and Tom Rosburg, "Fire in Central North American Grasslands: Vegetative Reproduction, Seed Germination, and Seedling Establishment," in Scott L. Collins and Linda L. Wallace, eds., *Fire in North American Tallgrass Prairies* (Norman: University of Oklahoma Press, 1990); T. J. Svejcar and J. A. Browning, "Growth and Gas Exchange of *Andropogon gerardii* as Influenced by Burning," *Jour. Range Management* 411 (1988) pp. 239–244.

68 Effects of fire on woody plants: S. R. Archer and L. L. Tieszen, "Plant

Response to Defoliation: Hierarchical Considerations," in *Grazing Research at Northern Latitudes*, ed. O. Gudmundsson (New York: Plenum, 1968).

68–69   Indian firing of the prairie: Stephen J. Pyne, *Fire in America: A Cultural History of Wildland and Rural Fire* (Princeton: Princeton University Press, 1982) pp. 84–85.

69   "The amazing herds of buffalo": John Filson, *The Discovery and Settlement of Kentucke* (Ann Arbor: University Microfilms, 1966) Facsimile of 1784 edition, p. 32.

70–71   Variations in grassland habitat: Paul G. Risser, "Landscape Processes and the Vegetation of the North American Grassland," in Collins and Wallace, eds., *Fire in North American Tallgrass Prairies*.

71   Raindrops on prairie ground surface: E. A. Fitzpatrick, *An Introduction to Soil Science* (Edinburgh: Oliver & Boyd, 1974) pp. 18–19.

71–72   Elk and mountain goats on the prairie: Russell McKee, *The Last West: A History of the Great Plains of North America* (New York: Crowell, 1974) p. 169.

72–73   Buffalo wallows: David J. Gibson, "Effect of Animal Disturbance on Tallgrass Prairie Vegetation," *Amer. Midland Naturalist* 121 (1988) pp. 144–154; and Frank Gilbert Roe, *The North American Buffalo: A Critical Study of the Species in Its Wild State* (Toronto: University of Toronto Press, 1951) p. 103.

73   Recharge ponds: T. C. Lee, A. E. Williams, and C. Wang, "Artificial Recharge Experiment in San Jacinto Basin, Riverside, Southern California," *Jour. Hydrology* 140 (1992) pp. 235–259.

73–74   Prairie dog population: Ernest Thompson Seton, in *Lives of Game Animals*, estimated that 5 billion prairie dogs lived on the nineteenth-century prairies; the U.S. Department of Agriculture, after twenty years of poisoning, estimated that in 1919 1.3 billion prairie dogs were inhabiting about 100 million acres of prairie. (For density estimates, see S. Archer, M. G. Garrett, and J. K. Detling, "Rates of Vegetation Change Associated with Prairie Dogs Grazing in North American Mixed-Grass Prairie," *Vegetation* 72 [1987] pp. 159–166.)

74   Prairie dog tunnels: R. G. Sheets, R. L. Linder, and R. B. Dahlgren, "Burrow Systems of Prairie Dogs in South Dakota," *Jour. Mammalia* 52 (1971) pp. 451–453. Also see David F. Costello, *The World of the Prairie Dog* (New York: Lippincott, 1970).

74   Preferential grazing of prairie dog towns: April D. Whicker and James K. Detling, "Ecological Consequences of Prairie Dog Disturbances," *BioScience* 38:11 (1988) pp. 778–784; and D. L. Coppock, J. K. Detling, J. E. Ellis, and M. I. Dyer, "Plant-Herbivore Interactions in a North American Mixed-Grass Prairie," *Oecologia* 56 (1983) pp. 1–15.

74   Temperature of prairie dog tunnels: Costello, *World of the Prairie Dog*, p. 86.

74–75   Prairie dog burrows as shelter for other animals: Ted Williams, "No Dogs Allowed," *Audubon*, September–October 1992, pp. 26–34.

75 Prairie dog's "orgy of infanticide": William K. Stevens, "Prairie Dog Colonies Bolster Life on the Plains," *New York Times*, July 11, 1995.

76 Prairie dog appetite: Leon H. Kelso, "Food Habits of Prairie Dogs," USDA Circ. 529, June 1939, pp. 1–15.

76 Macropore transport of water: E. L. McCoy, C. W. Boast, R. C. Stehouwer, and E. J. Kladivko, "Macropore Hydraulics: Taking a Sledgehammer to Classical Theory," in R. Lal and B. A. Stewart, eds., *Soil Processes and Water Quality* (Boca Raton: Lewis, 1994).

76 Water percolation in areas of high soil moisture: A. V. Granovsky, E. L. McCoy, W. A. Dick, M. J. Shipitalo, and W. M. Edwards, "Water and Chemical Transport through Long-Term No-Till and Plowed Soils," *Soil Sci. Soc. Amer. Jour.* 57 (1993) pp. 1560–1567. Also see Edwards, Shipitalo, Dick, and L. B. Owens, "Rainfall Intensity Affects Transport of Water and Chemicals through Macropores in No-Till Soil," *Soil Sci. Soc. Amer. Jour.* 56 (1992) pp. 52–58.

77 Buffalo shot for their tongues: Roe, *The North American Buffalo*, p. 342.

77 Nathaniel Henderson is quoted in ibid., p. 246.

77–78 Economics of the buffalo trade: William T. Hornaday, "The Extermination of the American Bison, with a Sketch of Its Discovery and Life History," *Smithsonian Report*, 1887 (Washington, D.C., 1889) Part II, p. 496.

78 John James Audubon: quoted in Alexander Adams, *Sunlight and Storm: The Great American Plains* (New York: Putnam's, 1977) pp. 256–257.

79 Scarcity of buffalo after 1849: Hornaday, "Extermination," p. 492.

79 Early cattle trails: David F. Costello, *The Prairie World* (New York: Crowell, 1969) p. 213.

79 Of the three railroads operating in the region at the time, two provided figures. These numbers are based on extrapolating an estimate from the nonreporting railroad. See Roe, *The North American Buffalo*, p. 437.

80 Effect of the Northern Pacific Railway on the northern herd: Ibid., p. 448.

80–81 Removal of the Indians: Rodman W. Paul, *The Far West and the Great Plains in Transition 1859–1900* (New York: Harper & Row, 1988) pp. 128–138.

81 British investors in cattle industry: McKee, *The Last West*, p. 240.

82–83 Advantages of grazing cattle in prairie dog towns: Whicker and Detling, "Ecological Consequences," p. 783; and D. M. Swift, "A Simulation Model of Energy and Nitrogen Balance for Free-Ranging Ungulates," *Jour. Wildlife Management* 47 (1983) pp. 620–645.

CHAPTER 6: PLOWING THE PLAINS

Page

86 Sod house construction: Alexander Adams, *Sunlight and Storm: The Great American Plains* (New York: Putnam's, 1977) p. 386.

87 Railroad rights on public land: Russell McKee, *The Last West: A History of the Great Plains of North America* (New York: Crowell, 1974) p. 228.

87 Requirements of the Timber Culture Act: R. Douglas Hurt, *The Dust Bowl: An Agricultural and Social History* (Chicago: Nelson-Hall, 1981), p. 19.

88 The Mennonites and the introduction of winter wheat: McKee, *The Last West*, pp. 260–262.

89 The hours of labor required to produce 100 bushels of wheat dropped from 233 man-hours in 1840 to 108 in 1900; 100 bushels of corn took 276 man-hours to produce in 1840 and 147 in 1900. Today 100-bushel quantities of those same grains take fewer than 9 and 7 man-hours respectively. U.S. Bureau of the Census, *Historical Statistics of the United States, Colonial Times to 1970* (Washington: U.S. Government Printing Office, 1975) p. 500.

89 The effects of the blizzards of 1886: Marc Reisner, *Cadillac Desert: The American West and Its Disappearing Water* (New York: Viking, 1986) pp. 109–110; and John Opie, *Ogallala: Water for a Dry Land* (Lincoln: University of Nebraska Press, 1993) p. 69.

89 The drought of 1887–1890: Opie, *Ogallala*, pp. 68–69.

89 The anecdote of the mule who froze to death in the heat is from Reisner, *Cadillac Desert*, p. 42.

89 "In God we trusted": McKee, *The Last West*, pp. 250–251.

90 The disk plow: Paul Bonnifield, *The Dust Bowl: Men, Dirt, and Depression* (Albuquerque: University of New Mexico Press, 1979) p. 52.

91 Increase in plowed acreage in Kansas in 1930s: Hurt, *Dust Bowl*, p. 99.

91 Effects in the East of prairie dust, 1934: Ibid., pp. 29–34.

91–92 The Kansan housewife is quoted in Donald Worster, *Dust Bowl: The Southern Plains in the 1930s* (New York: Oxford University Press, 1979) p. 17.

94 Advances in well technology: Donald E. Green, *Land of the Underground Rain* (Austin: University of Texas Press, 1973) pp. 49–61.

94 Output of the Ogallala in West Texas 1935–1950: William Ashworth, *Not Any Drop to Drink* (New York: Summit, 1982) p. 102.

94 Sixteen million acres irrigated by Ogallala today: Opie, *Ogallala*, p. 294.

95 Number of Mormon villages in Utah and Arizona in the nineteenth century: Rodman W. Paul, *The Far West and the Great Plains in Transition 1859–1900* (New York: Harper & Row, 1988) pp. 175–176.

96 "Crushed and mangled skeletons": Reisner, *Cadillac Desert*, p. 113.

96 "By 1889, there were only 5,700 square miles of irrigated cropland": Ibid., p. 116.

96–97 Results of 1894 Carey Act: Ibid., p. 114.

NOTES

CHAPTER 7: THE WATER OVER THE DAM

Page

101 The Roosevelt Dam and provisions of the Reclamation Act: Donald Worster, *Rivers of Empire* (Oxford: Oxford University Press, 1985) pp. 172–173.

102 Early reclamation projects: Brookings Institution, Institute for Government Research, *The U.S. Reclamation Service: Its History, Activities and Organization* (New York: Appleton, 1919) pp. 24–26.

102 Comparative figures for 1917 farm profits: Worster, *Rivers*, p. 179.

103 The Columbia River project as a 96.7-percent public subsidy: U.S. Dept of the Interior study cited in Bruce Brown, *Mountain in the Clouds: A Search for the Wild Salmon* (New York: Simon & Schuster, 1982) p. 84.

104 "Over a period of eighty-six years": This list of public reclamation projects is taken from U.S. Department of Interior, Bureau of Reclamation, *General Statistics for Calendar Year 1991*, p. 17.

105 Introduction of catfish and rainbow trout to Colorado River: Phillip L. Fradkin, *A River No More* (New York: Knopf, 1981) p. 242.

106 Geoffrey E. Petts, *Impounded Rivers: Perspectives for Ecological Management* (New York: Wiley, 1984) gives a full account of the changes in water quality and temperature in reservoirs and downstream from reservoirs.

107 Effect of stream temperatures on mayflies: Ibid., p. 197.

107 Early accounts of shad, herring, and eels: Erhard Rostlund, *Freshwater Fish and Fishing in Native North America* (Berkeley: University of California Press, 1952) pp. 14, 36.

108 Size of the Atlantic salmon: Brown, *Mountain in the Clouds*, p. 102.

109 Decreased salmon runs on Yakima and upper Columbia: Ibid., pp. 64–65, 85.

110 Obstacles faced by salmon heading up the Columbia River: Martin Heuvelmans, *The River Killers* (Harrisburg, Pa.: Stackpole, 1974) pp. 152–153.

111 Young salmon following underground watercourses: Brown, *Mountain in the Clouds*, p. 93.

111 Effects of reservoirs on salmon: W. J. Ebel, "Review of Effects of Environmental Degradation on the Freshwater Stages of Anadromous Fish," in John Alabaster, ed., *Habitat Modification and Freshwater Fisheries* (London: Butterworths, 1985) p. 70.

111 Smolt mortality rate of 95 percent: H. L. Raymond, "Migration Rates of Yearly Chinook Salmon in Relation to Flows and Impoundments on the Snake and Columbia Rivers," *Trans. Amer. Fisheries Soc.*, 97 (1968) pp. 356–359.

112 Fishkills from nitrogen supersaturation: Heuvelmans, *River Killers*, p. 158; and Petts, *Impounded Rivers*, pp. 228–229.

112 Beaver dams as salmon habitat: Hiram W. Li, Carl B. Schrenk, Carl E.

Bond, and Eric Rexstad, "Factors Influencing Changes in Fish Assemblages of Pacific Northwest Streams," in William Matthews and David C. Heins, eds., *Community and Evolutionary Ecology of North American Stream Fishes* (Norman: University of Oklahoma Press, 1987) p. 200.

112–13 Elwha hatchery and failure of fish farm program in Washington 1958–1966: Brown, *Mountain in the Clouds*, pp. 64–65, 85, 151–152.

113 Learning in hatchery fish: Terry DiVietti, Invited position address regarding the question "Hatchery and wild salmonids—can they coexist?" presented to Joint House Fisheries and Wildlife Committee of the State of Washington House of Representatives (June 1992).

113–14 Genetic variation in stream fish: Matthews and Heins, eds., *Community and Evolutionary Ecology*, p. 6.

CHAPTER 8: MUSSELS, GATORS, AND THE CORPS

Page
119 The Queen Pearl: Philip V. Scarpino, *Great River: An Environmental History of the Upper Mississippi, 1890–1950* (Columbia: University of Missouri Press, 1985) p. 81.

120 Nineteenth-century pearl hunting: Alexander Farn, *Pearls: Natural, Cultured and Imitation* (London: Butterworths Gem Books, 1986); and George Frederick Kunz, *Gems and Precious Stones of North America* (Chicago: Scientific Publishing, 1892).

121 Mississippi mussel harvest: Scarpino, *Great River*, p. 84.

121 Employment in the pearl-button business: John Sinkankas, *Gemstones of North America* (Princeton: Van Nostrand, 1959) p. 592; Scarpino, *Great River* pp. 94–95; and Farn, *Pearls*, p. 52.

122 Bureau of Fisheries' commercial propagation program: Scarpino, *Great River*, pp. 80–108.

122 Bureau of Fisheries' fish-rescue program: Ibid., pp. 103–105.

123 Medium-size Mississippi fish include mooneyes, bullheads, suckers, gizzard shad, skipjack, mud pickerel, white bass, black bass, crappies, and sunfish. With regard to the bottom of the carnivorous ladder, the country's most varied and abundant stock of minnows is found in the Mississippi River system. See Erhard Rostlund, *Freshwater Fish and Fishing in Native North America* (Berkeley: University of California Press, 1952).

124 Maintenance of open-water habitat by alligators: Constance E. Hunt, *Down by the River: The Impact of Federal Water Projects and Policies on Biological Diversity* (Washington: Island Press, 1988) p. 167.

125 Fish spawning in flooded Mississippi forests: Terry R. Finger and Elaine M. Stewart, "Response of Fishes to Flooding Regime in Lowland Hardwood Wetlands," in William Matthews and David C. Heins, eds., *Community and Evolutionary Ecology of North American Stream Fishes* (Norman: University of Oklahoma Press, 1987) p. 86.

127 Figures from Corps projects are from *Annual Report Fiscal Year 1992 of the Secretary of the Army on Civil Works Activities, Vol. 2* (Washington: U.S. Department of Defense, 1994) p. 1; see also *National Waterways Roundtable Proceedings: National Waterways Study, April 22–24, 1980* IWR–80-1 (Washington, D.C.: U.S. Government Printing Office, 1980) p. 652, and Martin Heuvelmans, *The River Killers* (Harrisburg, Pa.: Stackpole, 1974).

127 Corps stripping of phreatophytes: Heuvelmans, *River Killers*, pp. 136–138.

128 Tippah River study: Ibid., p. 206.

128–29 Locks and dams on the upper Mississippi: Scarpino, *Great River*, pp. 7–8, 41.

130 Dredging the lower Mississippi: Andrew Brookes, *Channelized Rivers: Perspectives for Ecological Management* (New York: Wiley, 1988) pp. 18–19.

131 Mussels as endangered species: John H. Cushman Jr., "Freshwater Mussels Facing Mass Extinction," *New York Times*, October 3, 1995; and National Research Council, *Restoration of Aquatic Ecosystems: Science, Technology and Public Policy* (Washington: National Academy Press, 1992) pp. 177–178.

CHAPTER 9: AQUEDUCTS AND TOILET BOWLS

Page

134 Water supply in ancient Rome: Clemens Herschel, *The Two Books on the Water Supply of the City of Rome of Sextus Frontinus* (Boston: Dana & Estes, 1899).

135 Hall is quoted in Charles Panati's *Extraordinary Endings of Practically Everything and Everybody* (New York: Harper & Row, 1989), p. 24.

135 Putresence in Boston's wells: Loammi Baldwin, *Report on Introducing Pure Water into the City of Boston* (Boston: Hilliard, Grey, 1835) p. 48; and Nelson Manfred Blake, *Water for the Cities* (Syracuse: Syracuse University Press, 1956) p. 13.

135 "The smell in many cases": Baldwin, *Report*, p. 77.

136 "A most subtle, peculiar . . . exhalation": W. Boghurst, *Loimographia, an Account of the Great Plague of London in the Year 1665*, ed. J. F. Payne (London, 1894).

136 Paris manure: Donald Reid, *Paris Sewers and Sewermen: Realities and Representations* (Cambridge: Harvard University Press, 1991) fn. p. 11.

137 Mary Gilliatt, *Bathrooms* (New York: Viking, 1971) p. 16.

138 *Annual Report of the Watering Committee*: noted in Blake, *Water*, p. 89.

138 Cochituate aqueduct: Ibid., p. 216.

139 John Melish, *Travels in the United States of America in the Years 1806 & 1807, and 1809, 1810 & 1811* (Philadelphia: Palmer, 1812) vol. 1, p. 61.

140 *Flushed with Pride: The Story of Thomas Crapper* (Englewood, N.J.: Prentice-Hall, 1971) by the British humorist Wallace Reyburn, purports to be a biography of the inventor of the flush toilet. This satirist also wrote a sly biography of the inventor of the bra, one Otto Titzling. Both are hoaxes, but Thomas Crapper, in particular, is working his way into the literature.

141 Water use in Boston hotels in the mid-nineteenth century: Blake, *Water*, p. 269.

141 Boston sewer statistics: Eliot C. Clarke, *Main Drainage Works of the City of Boston* (Boston: Rockwell & Churchill, 1888) p. 11.

142 Mortality rate from typhoid: Blake, *Water*, p. 260.

143–44 Filtering of water supplies: Allen Hazen, *The Filtration of Public Water Supplies* (New York: Wiley, 1896) p. 3.

144–45 Hazen Theorem: George Clifford White, *Handbook of Chlorination* (New York: Van Nostrand Reinhold, 1972) p. 285.

145 Early attitudes toward waste disposal in waterways: See, for example, George W. Fuller, "Is It Practicable to Discontinue the Emptying of Sewage into Streams?" *American City*, 7 (1912) pp. 43–45; and Ibid., "Relations between Sewage Disposal and Water Supply Are Changing," *Engineering News Record* 28 (1917) pp. 11–12.

146–47 Early industrial pollution: Joel Tarr, "Industrial Wastes and Public Health: Some Historical Notes, Part I, 1876–1932," *American Journal of Public Health*, Sept. 1985, Vol. 75, No. 9 pp. 1059–1067.

CHAPTER 10: DOWN THE DRAIN, UP THE STACK

Page

150 "Less than 1 billion pounds in 1940": *Historical Statistics of the United States, Colonial Times to 1970, Bicentennial Edition* (Washington D.C.: U.S. Bureau of the Census, 1975) compiled from Series P18–39 and P40–57.

150 "Today there are some 70,000 widely used synthetic chemicals": G. Tyler Miller, *Living in the Environment: An Introduction to Environmental Science*, 7th ed. (Belmont, Calif.: Wadsworth, 1992) p. 57.

150 My kitchen quarter-teaspoon holds between 36 and 40 drops. There are 3 teaspoons per tablespoon, 2 tablespoons per fluid ounce, and 128 fluid ounces per gallon. So a drop in 8.6 gallons is equal to a part per million, and 3 parts per trillion is a drop in 2,855,000 gallons.

150–51 Distribution of DDT in the food chain: Miller, *Living in the Environment*.

151 PCBs in electrical insulators: Laurent Hodges, *Environmental Pollution: A Survey Emphasizing Physical and Chemical Principles* (New York: Holt, Rinehart & Winston, 1973) p. 200.

151 Mercury use in manufacturing: "Mercury in the Environment," in Stanton S. Miller, ed., *Water Pollution: Articles from Volume 4–7 of Environmental Science and Technology* (Washington: Amer. Chem. Soc., 1974) pp. 253–255.

151–52 Minimata: W. Eugene Smith and Aileen M. Smith, *Minimata* (New York: Holt, Rinehart & Winston, 1975), pp. 26–33.

152 The presence of dioxins in papermaking effluent: Richard A. Bartlett, *Troubled Waters: Champion Paper and the Pigeon River Controversy* (Knoxville: University of Tennessee Press, 1995) pp. 212–213, 219.

152–53 The pioneering work of two British researchers, Dr. Richard M. Sharpe and Dr. John Sumpter, on the estrogen-mimicking effect of various industrial chemicals is well described in Lawrence Wright's "Silent Sperm," *The New Yorker,* January 15, 1996, pp. 47–51.

153 Estrogenic chemicals linked to reproductive disorders: Theodora Colburn, Frederick von Saal, and Ana Soto, "Developmental Effects of Endocrine Disrupting Chemicals in Wildlife and Humans," *Environmental Health Perspectives* 101:5 (October 1993) pp. 378–384.

153 Worldwide drop in human sperm count: See especially Elisabeth Carlsen, Aleksander Giwercman, Niels Keiding, and Niels Skakkebaek, "Evidence for decreasing quality of semen during past 50 years," *Brit. Med. Jour.,* September 12, 1992, pp. 609–612.

154 Pollution of the Calumet River: J. I. Bregman and Sergei LeNormand, *The Pollution Paradox* (Washington: Spartan Books, 1966) p. 54.

154 Wastes discharged by food-processing industries: Hodges, *Environmental Pollution,* pp. 166–167.

155 Amount of phosphates in city wastewater: Barry Commoner, *The Closing Circle* (New York: Knopf, 1971) p. 140.

156 Acid mine drainage in Ohio River Basin and Pennsylvania: Duane A. Smith, *Mining America: The Industry and the Environment, 1800–1980* (Lawrence: University Press of Kansas, 1987) pp. 114–115.

158 Current pesticide use in the United States: Jonathan Tolman, "Poisonous Runoff from Farm Subsidies," *Wall Street Journal,* September 8, 1995; Verlyn Klinkenborg, "A Farming Revolution: Sustainable Agriculture," *National Geographic,* December 1995, pp. 80–88.

158 Increase in nitrogen fertilizer on Illinois farms: Commoner, *Closing Circle,* p. 150.

159 Pollution in the Buffalo River: Noel M. Burns, *Erie: The Lake that Survived* (Totowa, N.J.: Rowman & Allanheld, 1985) p. 11.

159 The Cuyahoga as a fire hazard: Markham, *A Brief History of Pollution,* p. 62.

159 Reduced numbers of mayfly larvae in Lake Erie: William Ashworth, *The Late, Great Lakes: An Environmental History* (New York: Knopf, 1986) p. 124.

160 Death of beluga whales in the St. Lawrence estuary: Jon R. Luoma, "Doomed Canaries of Tadoussac," *Audubon,* March 1989, pp. 92–97.

162 Makeup of early environmental movement: Joseph Harry, Richard P. Gale, and John Hendee, "Conservation: An Upper Middle Class Movement," *Jour. Leisure Research*, Summer 1969; Ibid., "Conservation, Politics and Democracy," *Jour. Soil and Water Conservation*, November-December 1969, pp. 212–215.

CHAPTER I I : WHAT SLUDGE TELLS YOU

Page

167–68 1980s sewage in Boston Harbor: Berrin Tansel and Alice Outwater, "Boston Harbor Clean-Up," *Water, Environment & Technology* 3:3 (March 1991) pp. 45–50.

172 Binding of chemicals and heavy metals in sludge: Alice Outwater, *Reuse of Sludge and Minor Wastewater Residuals* (Boca Raton, Fla.: Lewis/CRC, 1994) p. 19.

174 U.S. Public Interest Research Group estimate: About 135,000 tons of industrial wastes were directly discharged to the waterways, while another 190,000 tons were released to municipal treatment plants. Assuming secondary treatment, about 10 percent, or 20,000 tons, would have made it into the waterways and oceans in 1992, for a total industrial contribution of 155,000 tons of toxins.

176–77 Critically endangered systems toward the smaller end of the scale include localized ecological communities such as the Hempstead Plains grasslands on Long Island, the wet coastal prairies in Louisiana, sedge meadows in Wisconsin, lake sand beaches in Vermont, native grasslands in California, and streams in the Mississippi alluvial plain.

177 Loss of riparian zone in Arizona and New Mexico: Roger L. DiSilvestro, *Reclaiming the Last Wild Places: A New Agenda for Biodiversity* (New York: Wiley, 1993) p. 116.

178 Increased trout habitat on Deerfield River: William K. Stevens, "New Rules for Old Dams Can Revive Rivers," *New York Times*, November 28, 1995.

179 1991 water flow regime on Colorado River: William Perry Pendley, *War on the West: Government Tyranny on America's Great Frontier* (Washington: Regnery, 1995) pp. 59–60.

179 Justice William O. Douglas on Corps: *Playboy* interview, July 1969.

180 Corps work on the Kissimmee and the Everglades: Mark Derr, "Reclaiming the Everglades," *Audubon*, September-October 1993, pp. 48–56; Martin Heuvelmans *The River Killers* (Harrisburg, Pa.: Stackpole, 1974) pp. 21–22; Richard Miniter, "Challenges Ahead for the EPA's New Earth Mother," *Insight*, February 8, 1993, pp. 6–11.

181 Zebra mussels: John Ross, "An Aquatic Invader Is Running Amok in U.S. Waterways," *Smithsonian*, February 1994, pp. 41–51.

183 Buffalo on public rangeland: Bureau of Land Management Programs, Fish and Wildlife Habitat Management, "Estimated Number of Big Game Animals on Public Lands, Fiscal Year 1991," in *The Executive Office's Council* *on Environmental Quality: 22nd Annual Report* (Washington: Govt. Printing Office, 1991) Table 25, p. 39.

184   Animal Damage Control Program: DiSilvestro, *Reclaiming*, p. 115.

184   Buffalo herds on private ranchland: Clifford D. May, "The Buffalo Returns, This Time as Dinner," *New York Times Magazine*, September 26, 1993, pp. 30–34.

184   Ted Turner's buffalo herd and return of prairie dog to his land: personal communication from Steve J. Dobrott, Turner's ranch manager.

185   Metapopulation theory: See especially Carol Kaesuk Yoon, "Wandering Butterflies May Be Charting the Path to Survival," *New York Times*, October 24, 1995.

186   Record 1995 bird migration south: William K. Stevens, "With Habitat Restored, Ducks in the Millions Create Fall Spectacle," *New York Times*, November 14, 1995.

INDEX

INDEX